Soil Forensics

Series editor
Henk Kars
Faculty of Earth and Life Sciences
VU University Amsterdam
Amsterdam, The Netherlands

To be a forum for all (scientific) workers in the rather fragmented field of Soil Forensics. This fragmented character is intrinsic to multidisciplinary research fields and a common platform for the exchange of knowledge and discussion is therefore heavily needed. To promote the field of Soil Forensics in academia, in forensic research institutes, legal profession/jurisdiction organisations and for the general public (science sections in newspapers) To contribute to a high scientific standard of the field. To be attractive for publishing in the series it is peer reviewed in order to be competitive with journals such as Forensic Science International.

More information about this series at http://www.springer.com/series/11807

Pier Matteo Barone • W. J. Mike Groen
Editors

Multidisciplinary Approaches to Forensic Archaeology

Topics discussed during the European Meetings on Forensic Archaeology (EMFA)

Springer

Editors

Pier Matteo Barone
Archaeology and Classics Program
American University of Rome
Rome, Italy

W. J. Mike Groen
Netherlands Forensic Institute
The Hague, The Netherlands

ISSN 2214-4293 ISSN 2214-4315 (electronic)
Soil Forensics
ISBN 978-3-030-06843-1 ISBN 978-3-319-94397-8 (eBook)
https://doi.org/10.1007/978-3-319-94397-8

This Springer imprint is published by the registered company Springer Nature Switzerland AG
The registered company address is: Gewerbestrasse 11, 6330 Cham, Switzerland

Preface

This volume contains papers presented during the last three European Meetings on Forensic Archaeology (EMFA) conferences. Some of these papers are written by researchers at the start of their carriers, other by senior and well-established academic scholars, forensic practitioners or military officers. The papers presented discuss case studies and deliberate on archaeological theory and methodology, forensic geophysics, human taphonomy, 3D-scanning, illegal trade of artefacts and forensic geology. Topics that are relevant to the advancement of forensic archaeology. This volume therefore offers a multidisciplinary view of forensic archaeology, though a personal perspective of the authors.

The first EMFA conference took place in 2012 in the Hague, the Netherlands, and was jointly organized by the forensic archaeology unit of the Netherlands Forensic Institute (NFI) and the forensic archaeology expert panel of the British Institute for Archaeologists (IfA). The purpose of the conference was to learn how forensic archaeology was organized and practised within different European countries, to meet foreign colleagues and to discuss the possibility of establishing a European forum for forensic archaeology. The conference offered 14 oral papers and was attended by circa 50 delegates, covering 12 countries and representing forensic science institutes, police forces, humanitarian organizations, academic institutions and freelance professionals. An additional outcome was the creation of an official platform for Forensic Archaeology in Europe, established within the European Network of Forensic Science Institutes (ENFSI) in 2013 (Groen 2015). Subsequent conferences were organized by the NFI in 2013 (The Hague, The Netherlands), the Cranfield Forensic Institute in 2014 (Shrivenham, UK), the Forensic Sciences Institute of the French Gendarmerie (IRCGN) in 2015 (Pontoise, France), the Independent Commission for the Location of Victims' Remains (ICLVR) and the Forensic Service Ireland (FSI) in 2016 (Dublin, Ireland) and the American University of Rome in 2017 (Rome, Italy). In total more than 120 oral papers and posters were presented, discussing case studies, field methodology, human taphonomy, forensic geophysics and quality assurances, to name a few topics. The forthcoming EMFA conference will be organized by the Queens University in Belfast, UK, and is scheduled for August 2018.

During the EMFA 2017 meeting in Rome, Italy, Pier Matteo Barone and Mike Groen, this volume's editors, discussed the possibility to publish oral papers, as presented during the last three EMFA conferences in the Soil Forensic series, edited by professor Henk Kars, as a third volume so far. As a result, a number of potential authors was contacted. Most of the contacted individuals kindly agreed to participate and submitted a paper for publication. It is hoped that this volume will not only improve the image of forensic archaeology worldwide, by producing more awareness about this discipline, but that it will also contribute to the theoretical and methodological development and a European 'Best Practice Manual' on forensic archaeology. It is also hoped that this volume and the EMFA initiative will improve the levels of communication and understanding between archaeologists, forensic scientists, police officers and other individuals who are – or potentially could be – involved in forensic archaeological research and/or case work. Moreover, to promote both the group identity and the communication, the editors of this book have decided to present officially the brand-new logo (Fig. 1) that will be the symbol of EMFA for now on.

This volume contains 18 chapters subdivided in three parts. Chapter 1, written by Mike Groen, is an introductory chapter and discusses the theoretical concepts behind the practice of forensic archaeology, focusing on domestic police cases. By doing so it offers an interpretative framework to evaluate physical evidence, by combining knowledge from archaeology, criminalistics and criminology. Part I, Forensic Archaeology and the NDT's, focuses on the theoretical and methodological issues in GIS-research, remote sensing and forensic geophysics. Chapter 2, written by Alastair Vannan, discusses the importance of geospatial analyses while preparing a search for buried persons or objects. By doing so it focuses on the relevance of LiDAR and the Red-Amber-Green (RAG) mapping to analyse and prioritize possible search areas. Chapter 3, written by Pier Matteo Barone, discusses a combined approach using forensic geophysics, cadaver dogs and forensic archaeology to detect buried persons or objects at crime scenes. It therefore offers a valuable multidisciplinary perspective on search methodology in forensic setting. Chapter 4, written by Alastair Ruffell and Laurance Donnelly, discusses the theory, methodology and usability of forensic geophysics, focusing mainly on the ground penetrating radar (GPR), electro-magnetics (EM) and magnetometry, inside human-made

Fig. 1 The official EMFA logo

structures, within peat bogs and (using a floating GPR, or water-penetrating radar) waterbodies. Part II, Forensic Archaeology and Human Remains, focuses on human taphonomical research and case studies involving buried persons. Chapter 5, written by Nicholas Márquez-Grant, discusses the role of a forensic anthropologist during a forensic search where a (partially) skeletonized body is expected to be encountered. Chapter 6, written by Patrice Georges, Christelle Buton and Éric Crépin, discusses the structure and practice of forensic archaeology in France, focusing on the Forensic Sciences Institute of the French National Gendarmerie (IRCGN). Chapter 7, written by Hayley Mickleburgh, discusses the usability of archaeological funerary taphonomy and actualistic experiments involving human inhumations in archaeology and forensic sciences. Chapter 8, written by Patrick Randolph-Quinney, Stephen Haines and Ashley Kruger, discusses the use of 3D-scanning for recording forensic taphonomic traces on bones, by focusing on the available technology, the visualization potential and validation of the obtained 3D-models. Chapter 9, written by Agathe Ribéreau-Gayon, Carolyn Rando and Ruth Morgan, discusses the challenges and future developments in human decomposition in deep water. By doing so it also focuses on forensic detection, recovery and identification of human remains in marine environments. Chapter 10, written by Tomasz Borkowski and Maciej Trzciński, discusses the issues associated with Polish forensic archaeology, by focusing on case study involving the search for buried remains of soldiers of a Polish underground unit who were killed in September 1946. Chapter 11, written by Lars Krants, discusses the structure and practice of forensic archaeology in Denmark, focusing on the collaboration between the Danish National Police and the Moesgaard Museum. Chapter 12, written by Chantal Milani and Carlo Belardo, discusses the structure and practice of forensic archaeology and anthropology in Italy, focusing on the Scientific Investigation Department of the Italian Carabinieri in Rome (RIS-Rome). Chapter 13, written by Tatyana Shvedchikova, discusses the development of forensic archaeology in the Russian Federation by focusing on the Russian research in human taphonomy and WWII graves. Chapter 14, written by Geert Jonker, presents a case study involving the excavation, analysis and identification of a British soldier who has been killed south of Arnhem, the Netherlands, in early October 1944, in the aftermath of Operation Market-Garden. Chapter 15, written by Geoffrey Knupfer, Dennis Godfrey and Jon Hill, discusses a case of the Independent Commission for the Location of Victims' Remains (ICLVR) involving the victim of the 'Troubles' who has been abducted, killed and secretly buried by republican paramilitaries in April 1978. Part III, Forensic Archaeology and Antiquity Crimes, focuses on the destruction, theft and illegal trade of archaeological and cultural artefacts. Chapter 16, written by Pier Matteo Barone, discusses the role forensic archaeology can play in solving crimes against cultural heritage, focusing on the Italian structure and examples. Chapter 17, written by Maciej Trzciński, discusses the current Polish situation in relation to investigation, prosecution and prevention of crimes against archaeological heritage. The last chapter, Chap. 18, is written by Rosa Maria Di Maggio and focusses on the role of forensic geology in forensic investigation related to fossil, precious gems and metals.

Finally, the editors would like to express their sincere gratitude to all contributors who participated in this volume. Without all of you this volume would not have been possible. We are also indebted to professor Henk Kars for offering us the opportunity to compile this volume and to the publishing team at Springer, in particular Melanie van Overbeek.

The Hague, The Netherlands
Rome, Italy
April 30th 2018

W. J. Mike Groen
Pier Matteo Barone

Contents

Contributors

Pier Matteo Barone American University of Rome, Rome, Italy

Geoscienze Forensi Italia® – Forensic Geoscience Italy, Rome, Italy

Carlo Belardo Crime Scene Investigation Unit – Carabinieri R.I.S. Scientific Investigation Department, Rome, Italy

Tomasz Borkowski Institute of National Remembrance, Warsaw, Poland

Christelle Buton French National Gendarmerie, Toulouse, France

Éric Crépin Paratroopers Forces (French Ministry of the Armed Forces), Paris, France

Rosa Maria Di Maggio Geoscienze Forensi Italia® – Forensic Geoscience Italy, Rome, Italy

Laurance Donnelly International Union of Geological Sciences, Initiative on Forensic Geology, Lancashire, UK

Patrice Georges French National Institute for Preventive Archaeological Research (Institut national de recherches archéologiques preventives; Inrap), Paris, France

Joint Research Unit TRACES ("Archaeological Works and Research on Cultures, Spaces and Societies"; UMR 5608), Toulouse University – Jean Jaurès (UT2J), Toulouse, France

Dennis Godfrey Independent Commission for the Location of Victims Remains (ICLVR), Dublin, Ireland

W. J. Mike Groen Netherlands Forensic Institute, The Hague, The Netherlands

Faculty of Archaeology, Leiden University, Leiden, The Netherlands

Stephen D. Haines School of Forensic and Applied Sciences, University of Central Lancashire, Preston, Lancashire, UK

Jon Hill Independent Commission for the Location of Victims Remains (ICLVR), Dublin, Ireland

Capt. Geert Jonker The Recovery & Identification Unit of the Royal Netherlands Army (RIU), Soesterberg, The Netherlands

Geoffrey Knupfer Independent Commission for the Location of Victims Remains (ICLVR), Dublin, Ireland

Lars Krants Moesgaard Museum, Højbjerg, Denmark

Ashley Kruger Evolutionary Studies Institute, University of the Witwatersrand, Johannesburg, South Africa

Nicholas Márquez-Grant Cranfield Forensic Institute, Cranfield University, Defence Academy of the United Kingdom, Shrivenham, UK

Hayley L. Mickleburgh Faculty of Archaeology, Leiden University, Leiden, The Netherlands

Chantal Milani Carabinieri R.I.S. Scientific Investigation Department, Rome, Italy

Ruth M. Morgan Department of Security and Crime Science, University College London, London, UK

Centre for the Forensic Sciences, University College London, London, UK

Carolyn Rando Institute of Archaeology, University College London, London, UK

Patrick S. Randolph-Quinney School of Forensic and Applied Sciences, University of Central Lancashire, Preston, Lancashire, UK

Evolutionary Studies Institute, University of the Witwatersrand, Johannesburg, South Africa

School of Anatomical Sciences, Faculty of Health Sciences, University of the Witwatersrand Medical School, Johannesburg, South Africa

Agathe Ribéreau-Gayon Department of Security and Crime Science, University College London, London, UK

Centre for the Forensic Sciences, University College London, London, UK

Institute of Archaeology, University College London, London, UK

Alastair Ruffell School of the Natural Built Environment, Queen's University, Belfast, N. Ireland, UK

Tatyana Shvedchikova The Institute of Archaeology Russian Academy of Sciences, Moscow, Russia

Maciej Trzciński Faculty of Law, Administration and Economics, Department of Forensic Sciences, University of Wrocław, Wrocław, Poland

Alastair Vannan Cellmark Forensic Services, Chorley, UK

Chapter 1
Forensic Archaeology: Integrating Archaeology with Criminalistics and Criminology

W. J. Mike Groen

Abstract This chapter suggests a forensic archaeological framework that integrates the archaeological process and research cycle with criminalistic and criminological knowledge. By doing so, it proposes evidential reasoning that is based on the Bayesian inference, physical evidence recovered at crime scenes, middle-range theories, site formation processes, forensic assemblages and mutually exclusive hypotheses. This to analyse, interpret and reconstruct the human (criminal) behaviour and natural processes that have formed and modified the investigated crime scene. The analysis and interpretation of the recovered forensic record should commence during field work, at the crime scene. After completing this investigation, the secured physical evidence, as analysed and interpreted by different fields of forensic expertise, should be integrated into one overall synthesis. In addition, a forensic archaeological investigation should not be an end in itself. It should also be used to generate inductive and deductive theories and models needed for reconstructing the human (criminal) behaviour and natural processes responsible for the formation, distortion and destruction of the physical evidence. To be able to do this, feedback from casework is essential, as are development and innovation. To strengthen its base as a scientific discipline, the chapter closes with a discussion on innovation and development within the field of forensic archaeology.

Keywords Forensic archaeology · Archaeological research cycle · Criminalistics · Criminology · Bayesian inference

---

W. J. M. Groen (✉)
Netherlands Forensic Institute, The Hague, The Netherlands

Faculty of Archaeology, Leiden University, Leiden, The Netherlands
e-mail: m.groen@nfi.minvenj.nl

P. M. Barone, W. J. M. Groen (eds.), *Multidisciplinary Approaches to Forensic Archaeology*, Soil Forensics, https://doi.org/10.1007/978-3-319-94397-8_1

1.1 Introduction

Forensic archaeology is habitually defined as the use of archaeological methods and techniques in a legal setting (e.g. Groen et al. 2015). Accordingly, forensic archaeological literature has a tendency to focus on the application of archaeological field methodology in forensic and/or medicolegal settings. The archaeological mainstream discourse on theory during the past decades (e.g. Bentley et al. 2009; Johnson 2010; Harris and Cipolla 2017) seems to have been mostly ignored within forensic archaeology although there are noteworthy exceptions (e.g. Crossland 2013; Gupta 2013; Fondebrider 2016). In addition, most forensic archaeological advancements seem to be methodical in nature, focusing, for example, on geographic information systems (GIS), remote sensing and geophysics. The study of taphonomic changes seems to be an exception to the above-mentioned. This focus on (geo- and medical) sciences could be problematic in the long run, since one of the outcomes of the archaeological discourse was that the analysis and interpretation of the archaeological record, needed to reconstruct past human activities, cannot be based on methodological research alone (e.g. Cleland 2002, 2011; Jeffares 2008). To link the empirical observations to inductive or deductive theories and models that are able to clarify the studied past human activities, archaeological reconstructions should combine empirical observations with an interpretative framework. In this way archaeology integrates concepts from the humanities (e.g. history), the social sciences (e.g. economics, cultural anthropology) and the sciences (e.g. biology, earth sciences) into its paradigm and is thus research data- and theory-driven. Since forensic archaeology is a form of (applied) archaeology, this framework should also be integrated with forensic archaeology.

Although the usability of the archaeological framework in crime scene investigation has already been discussed (Groen and Berger 2017), it has not – to author's knowledge – been successfully incorporated in forensic archaeological casework, as related to police cases (e.g. searches, excavations, exhumations and field recoveries). This chapter focuses on the applicability of the archaeological framework and research cycle to forensic archaeological cases combined with theory and methodologies from criminalistics and criminology. By doing so, it builds on the probabilistic framework as proposed by the author and colleagues in the concluding remarks of Groen et al. (2015). The chapter itself is divided into four parts. Part one briefly discusses the key concepts in archaeology, criminalistics and criminology. Part two introduces the archaeological research cycle and utilises this cycle to structure the forensic archaeological investigation. Part three focuses on development and innovation needed to aid forensic archaeology in improving its processes, probabilistic framework and investigative methodology. In conclusion part four summarises the main points of the chapter.

1.2 Archaeology, Criminalistics and Criminology

Archaeology can be defined as the study and reconstruction of past human activities and related natural processes, through a systematic recovery, analysis, interpretation and reconstruction of the archaeological record. The archaeological record covers human remains, material culture (artefacts and ecofacts), non-portable archaeological features and environmental information. The archaeological record is always recorded in situ and in context or in its original position and association. Archaeological field methodology is research question driven and developed to systematically search for and reliably document the archaeological record (e.g. Hester et al. 2008). Modern field recording systems combine spatial databases with 3D recording strategies (e.g. Smith et al. 2015) and unmanned aerial vehicles (UAVs) with structure from motion (SfM) photogrammetry (e.g. Howland et al. 2015) and are fully integrated with GIS applications. Essential for understanding the archaeological record is its context, association and the site formation process or all human and non-human (natural) processes that have formed and modified an archaeological site (e.g. Clarke 1973; Schiffer 1976, 1983, 1987). Site formation processes can be subdivided into non-human natural processes (n-transforms) and human natural processes (c-transforms). Since site formation processes could also be additive, they can introduce new features into the archaeological record. An archaeological reconstruction of the past must therefore not only focus on the retrieved archaeological record but must also take into account the site formation processes that (could) have modified, destroyed or masked the record that was originally deposited at the research site. Since archaeologists reason retrospectively when studying past societies, they use background knowledge and an interpretative framework to link the obtained archaeological record to inductive and deductive theories and models about the past.[1] Archaeological background knowledge is mostly obtained from (spatial) databases, theoretical models, material studies and actualistic experiments.[2] Archaeological databases are used to structure archaeological research, for example, prior to the excavation or recovery (as background knowledge), during the excavation or recovery (as recording strategy) and after the field work (as a reconstruction tool). Archaeological theoretical models are often based on middle-range theories, or MR-theories (Binford 1977, 1981; Schiffer 1976, 1988). The MR-theories are a set of testable (and falsifiable) models or propositions about human activity that link the spatio-temporal regularities in the archaeological record (the empirical domain) to the more abstract high-level conceptual schemes

[1] Archaeologists use different theoretical frameworks to analyse and reconstruct the past, depending on their epistemological and ontological viewpoints (e.g. Bentley et al. 2009; Johnson 2010; Harris and Cipolla 2017).

[2] For inferences about archaeological reasoning, see, for example, Sullivan (1978), Gibbon (1989), Wylie (2002, 2011), Alberti et al. (2013) or Chapman and Wylie (2015, 2016).

or theories about human behaviour (the theoretical domain).[3] Archaeological MR-theories could be constructed from actualistic experimental studies, historical sources, taphonomic research or ethnoarchaeological observations. Other MR-theories are abstracted from present day economic, sociologic or cultural anthropological theories and models. The above-mentioned means that an archaeological interpretation is not only data-driven but also theory-driven. In other words, the archaeological collection, analysis and interpretation of data are subjected to a selection process and are therefore interpretation driven. There is no such thing as objective, theory-free data collection or sampling.

Criminalistics can be defined as the scientific study concerning the recovery, interpretation, evaluation and reporting of physical evidence obtained in criminal cases, including the role of the forensic scientists (e.g. Saferstein 2015). For the prediction of the forensic value of the obtained physical evidence, Locard's exchange principle[4] and the principle of the 'evidentiary triangle' are used. This is in order to predict which item or trace could be linked to the victim, the suspect and the crime. Items or traces that can be linked to all three are, generally speaking, forensically most relevant since they can possibly be used to identify the (unknown) offender. As a consequence, criminalistic methodology focusses on the possible transfer, or lack thereof, and degradation of crime-related physical evidence. Criminalistic reasoning is based on the Bayesian approach, meaning that the evidential value of a forensic trace (to be usable in court) not only depends on the trace itself, the context of the criminal case and the location where this trace has been found but also on the hypotheses used to evaluate this trace. A hypothesis could be seen as the translation of a relevant forensic question into a realistically possible answer. The criminalistic evaluative framework is therefore probabilistic in nature and works with at least two mutually exclusive hypotheses (or a pair of propositions) and with comparison as the method of examination (e.g. Cook et al. 1998a, b; Evett et al. 2000; Evett 2015; Robertson et al. 2016). A suitable hypothesis that does well for explaining the observed results can thus be integrated into an overall police scenario concerning what (could have) happened at the crime scene. Since the prior odds of these hypotheses are considered to be the task of the police, or the court, the aim is to provide evidence to support, or falsify, one (or more) of the hypotheses by considering the probability of the evidence. The aim is not to test the probability of the hypotheses themselves.[5] Hypotheses, inferences and conclusions concerning the (transfer of) physical evidence can be investigated at the source level, at the activity level and at

[3] For critical discussion on the usability of MR-theories in archaeology, see, for example, Schiffer (1976, 1988), Binford (1977, 1981), Raab and Goodyear (1984), Pierce (1989), Wylie (1989), Salmon (1992), Maschner (1996), Tschauner (1996), Shott (1998) or Smith (2011).

[4] Locard's exchange principle states that the perpetrator of a crime could leave a physical trace on a crime scene, or bring a trace into it and that this trace could be used as forensic evidence (Locard 1920, 1930).

[5] For more information concerning the way (European) forensic laboratories value evidence and report conclusions, see the ENFSI 'Guideline for Evaluative Reporting in Forensic Science' (ENFSI 2015).

the level of the offence (e.g. Cook et al. 1998a; Evett et al. 2000).[6] In view of the aforementioned, forensic archaeological reasoning ought to comply with criminalistic reasoning and should therefore follow the Bayesian approach when evaluating physical evidence. In addition, forensic archaeological conclusions should be, if possible, formed at the activity level, since an explanation at the source level alone will not tell us how the retrieved evidence came to be deposited at the crime scene.

Criminology is the scientific study of crime, criminality, criminals, criminal behaviour and the criminal justice and penal systems (e.g. Frailing and Harper 2015). Of importance here is that criminology also studies the causality of crime and criminal behaviour and that GIS applications had been used to map and analyse crime distribution and patterns (e.g. Paulsen and Robinson 2008). For example, environmental criminology studies crime patterns within different environments and architectural features and the impact of these patterns on humans (e.g. Bruinsma and Johnson 2018). Another criminological discipline, offender profiling, utilises information on geographical locations (e.g. the location where the victim lived and the location where the victim was last seen), the spatial layout of crime scenes, the type of crime, the psychological theories concerning the offender and the personal history of the victim to predict the characteristics and place of residence of the offender in order to apprehend this individual (e.g. Rossmo 2000; Ainsworth 2001; Bartol and Bartol 2012). A number of offender profiling studies include locations of body deposition sites – as related to the type of crime, the type of victim and the type of offender – in their research framework[7] and are therefore highly relevant to forensic archaeology. This since such studies can be utilised to build MR-theories about where offenders hide their victims and can provide information that could help structure forensic archaeological search strategies involving missing persons. Moreover, criminological studies can offer (emic) interpretations about criminal behaviour and the spatial layout of physical evidence at crime scenes, as offered by the offenders themselves. In doing so, criminology can provide data and information needed to build additional MR-theories that could improve our understanding of the retrieved physical evidence. A careful approach is needed when analysing offenders' testimonies since they often do not give an entirely honest account of past events. However, given the fact that most of the academically trained forensic archaeologists are also accustomed to work within the social sciences framework, these limitations are not unfamiliar to forensic archaeologists.

[6] The source level relates to the identification or classification of the physical evidence in relation to its possible source. The activity level evaluates the relationship between the transferred material and the human action that possibly resulted in this transfer. The offence level focuses on the relationship between the transferred material and the offence.

[7] See, for example, Rossmo (2000), Lundrigan and Canter (2001a, b), Morton and Lord (2002), Nethery (2004), Snook et al. (2005), Powell (2006), Häkkänen et al. (2007), Van Patten and Delhauer (2007), Hays (2008) or Anderson (2009).

1.3 The Archaeological Research Cycle

Given the above-mentioned, a forensic archaeological site should be seen as an accumulation of materialized patterns of past human and non-human activities, often distorted. Additionally, we should combine our methodology with criminalistics and attempt to develop a probabilistic framework – based on the Bayesian inference – to reconstruct the human activities and natural processes that took place at the crime scene (prior to, during and after the crime has been committed). To be able to do this, we also need to build forensically relevant MR-theories that are able to link the retrieved forensic record (the sum of all archaeologically retrieved physical evidence) with inductive and deductive theories and models about the behaviour that initially formed this record. Additionally, we need to restructure forensic archaeological investigations in accordance with the archaeological research cycle. This cycle covers (after Swartz 1967) a sequence of nine steps: (1) preliminary desktop research, (2) field survey, (3) final desktop research, (4) excavation or recovery, (5) data analysis, (6) interpretation, (7) integration, (8) comparison and (9) abstraction. The time frame of each step varies per archaeological site since some excavation campaigns take place annually and can continue for decades. However, most research into archaeological sites will complete this cycle within a year or so. The sequence of steps in the cycle is not set in stone; it is possible to retreat one or more steps, for example, when unexpected physical evidence is located or new questions are put forward. Since archaeologists initiate data analysis and interpretation during fieldwork, some of these steps are, also, interwoven with one another, especially steps 4 (excavation or recovery), 5 (data analysis) and 6 (interpretation).

The remaining part of this section attempts to structure a forensic archaeological investigation according to the nine archaeological research cycle steps, starting with preliminary desktop research and ending with abstraction of the obtained forensic record.[8]

1.3.1 Preliminary Desktop Research

The preliminary desktop research (or desk-based assessment (DBA)) is the backbone of all archaeological research. By accessing information acquired from previous research (the background knowledge), archaeologists are able to express an expectation of what an forthcoming excavation would bring, which deductive research hypotheses can be addressed and which scientific questions could be answered. These expectations also lead to the development of a preliminary excavation or recovery strategy and field methodology.

[8] Since the discussion of methods and techniques used during the crime scene investigation falls outside the scope of this chapter, the reader is forward to Fisher and Fisher (2012), Gardner (2012) or Houck et al. (2012). For discussion of forensic archaeological field methodology, see, for example, Dupras et al. (2011), Hunter et al. (2013), Dirkmaat and Cabo (2016) or Evis (2016).

Integrating preliminary desktop research into forensic archaeological investigation would surely benefit such investigation. By assessing background knowledge, for example, a GIS application with pedological, geomorphological and ecological maps, the forensic archaeologist would be able to predict what the soil and ecological conditions are at and near the to be investigated crime scene, or place of incident. This information would enable the archaeologist to estimate which formation processes and taphonomic transformations (e.g. taphonomic changes, soil and ecological alterations and material degradation) could have taken place at this site and, therefore, to predict which methods and techniques would offer the best change to find a body, or object. The above-mentioned could be 'easily' done within a Bayesian framework of reasoning, for example, by providing prior odds about the detectability of a grave in different pedological, geomorphological and ecological contexts by means of field walking, geophysics, remote sensing or cadaver dogs.

Additionally, background knowledge obtained from criminological MR-theories concerning human behaviour, case review studies and experimental studies would aid the development of deductive forensic hypotheses that should be addressed during the investigation. Given the type of investigated crime, for example, a surface scatter scene related to a rape and subsequent murder or a clandestine burial related to a murder in a domestic abuse context, this offers a probabilistic framework to reconstruct the activities that took place at the crime scene. Assessing background knowledge, by means of archaeological predictive modelling (e.g. Verhagen and Whitley 2012; Lieskovský et al. 2013), would also inform the forensic archaeologist about the type of physical material that could be expected, where this material might be located and what the evidentiary value of this material could be. Thus prior odds concerning the spatial layout of the forensic record, the deductive hypotheses of interest and the activities that were responsible for the spatial distribution of this record at a given crime scene can be estimated. In this way, the preliminary desktop research could be compared with the pre-assessment phase of the criminalistic framework (e.g. Cook et al. 1999). The prior odds, if used during the crime scene investigation phase, will optimise the investigative methodology and therefore the chance of collecting the most relevant physical evidence (Groen and Berger 2017). However, using background knowledge and ascribing prior odds before and during the crime scene investigation could also lead to contextual biases (e.g. Ditrich 2015; Stoel et al. 2014). In addition, confirmation bias (e.g. Kassin et al. 2013) should be kept in mind to avoid a situation where only physical evidence that supports the investigated deductive hypotheses is collected. Investigations should therefore also collect evidence that could falsify the stated forensic hypotheses.

In goes without saying that introducing preliminary desktop research in forensic archaeology will cause difficulties given the time restraints of a forensic case. Forensic archaeologists working on police cases do not have the time to perform an extensive preliminary desktop research, barring some cold cases. However, this does not mean that we should omit this step in forensic archaeology. As discussed in the development and innovation section, it all depends on configuring the background knowledge as indexed documents or spatial databases.

1.3.2 Field Survey

In forensic archaeology a field survey, if backed with the knowledge obtained from the preliminary desktop research, will allow a limited testing of forensic hypotheses and could be helpful for the improvement of the investigative strategy and methodology. Forensic archaeological field surveys generally consist of systematic surface walkover, geophysics, remote sensing and/or subsurface testing in the vicinity of the crime scene or place of incident, to examine local soil features, for example, by probing, auguring or test pits. The available background knowledge and prior odds, as obtained by means of the preliminary desktop research, should subsequently be updated during and after the field survey. This should be done by inspecting the crime scene, or place of incident, and by obtaining case-specific information during a police briefing.

Additionally, a forensic archaeological field survey should be research and question driven, integrating 3D mapping, GIS and database features into its field procedures. During the survey, the crime scene should be assessed for the potential availability of (different types of) forensic materials, given the type of crime investigated, the locally encountered conditions and the time frame of the case. In this way, a field survey can be to a certain extent compared with the assessment phase of the criminalistic framework (e.g. Cook et al. 1998b). Additionally, preliminary information concerning the site formation processes and taphonomic transformations should be collected and processed. For example, to predict which types of material evidence are, or could be, no longer present at the scene, for instance, due to taphonomic changes or degradation. Therefore assessment will not only improve the knowledge obtained from the preliminary desktop research but will also update during this research obtained prior odds and therefore the investigative strategy, forensic research questions and hypotheses that need to be considered during the investigation. For example, the detectability of a grave will differ in different soil conditions and be dependent on the taphonomic conservation of the to be located human remains. Therefore a negative result given a search tool, for example, the ground-penetrating radar (GPR), does not have to mean that the to be located body is not present within the searched area. This result still will need to be evaluated using the Bayesian inference. In other words, field surveys are not only important for orientation and location of potential physical evidence. They are also important for the updating of evidential value of the (to be expected) physical evidence and are therefore also relevant in inductive hypotheses and scenario building.

1.3.3 Final Desktop Research

Given the above-mentioned, final desktop research should aim to formulate the final forensic research questions and hypotheses and formulate the investigative strategy and methodology, using the Bayesian framework of reasoning. This should ideally be done back at the office, after finalising the field survey. However, this could also be done at the crime scene, following the field survey and before engaging in the excavation or recovery. It goes without saying that the formulated investigative strategy and methodology should be in agreement with the to be answered forensic questions. The employed methodology should be able to utilise theories and models enabling the forensic archaeologist to recognise, analyse and interpret the forensic record, not only after but also during the investigation of the crime scene. The investigative methodology should furthermore be case specific, be forensic hypothesis driven, be able to optimise the allocation of search efforts and be able to maximise the retrieval of evidence with the highest evidential value, as related to the formulated hypotheses. The prediction of the type and size of the physical evidence that could have been exchanged between the scene and the victim, between the victim and perpetrator and between the scene and the perpetrator should also be conducted. The formulated investigative methodology should additionally be adjusted if unexpected physical evidence is encountered or if the stated forensic hypotheses are falsified. Finally, all methods employed will have to account for limitations that could impact the retrieved forensic record, such as terrain and light conditions, the size of the to be investigated crime scene and the available manpower, experience and time.

Since the documentation, analysis and interpretation of the forensic record should be centred around – in archaeological terms – assemblages or a combination of different types of in situ physical evidence occurring in a defined space (e.g. Lucas 2012), it is important to recognise the events and processes that have created and subsequently affected the to be investigated crime scene. The formulated investigative methodology should therefore be able to document the encountered stratigraphy and contextual association in three dimensions and be able to recognise the local site formation processes and taphonomic transformations.

1.3.4 Excavation or Recovery

A forensic excavation or recovery concentrates on the recognition, documentation, collection and processing of the forensic record. The main goals, from the forensic archaeological point of view, are to gather evidence that can be used to support or falsify hypotheses, to build scenarios and to develop a crime scene formation model that can help reconstruct the human activities and natural processes that took place at the crime scene, prior to, during and after the crime has been committed (see next sections). Each forensic archaeological investigation should therefore follow archaeological and criminalistic principles.[9] As discussed earlier; predictive models

[9] Examples of archaeological principles are the principles of superposition, original horizontality and lateral continuity. Examples of criminalistic principles are the Locard's exchange principle and the principle of the 'evidentiary triangle'.

and MR-theories should be utilised to inform the forensic archaeologist as to where new physical evidence could be expected not only based on what has been found so far but also on the available background knowledge. In this way the investigation will be hypothesis driven and will make use of abductive reasoning to proceed. However, what type of physical evidence that will be collected and how this is done (using standard operating procedures or SOPs) will depend on the specific national protocols and jurisdictions and the limitation of the forensic laboratory tests (i.e. false positives and false negatives).

It is a truism that the forensic value of material evidence studied at a forensic laboratory is only as good as its recognition, documentation and recovery at the crime scene. Since collection, analysis, interpretation and evaluation of physical evidence goes hand in hand; evidence collection and evidence evaluation are interdependent rather than independent. Collection and sampling of physical evidence are therefore subject to a selection process, based on the theoretical and methodological aspects and the background knowledge available to the investigator. Each forensic archaeological investigation should consequently be site plan led. Because excavation and recovery destroys the spatial associations within an assemblage site, plans should be created on site, preferably using objective documentation techniques, for example, a total station, structure from motion (SfM) photogrammetry, ortho-photogrammetry and/or 3D scanning. If not recorded properly, the destructive nature of an archaeological investigation will lead to loss of evidence at best, but it could also lead to ambiguous or erroneous interpretation of the obtained forensic record. Consequently, mapping should not be seen as a documentation tool only but also an assessment tool, especially if this assessment is performed within a GIS application.[10] Digital mapping of the forensic record will not only improve the accuracy of documentation, interpretation, reconstruction and visualisation of this record but will also save time and improve the chain of custody.

1.3.5 Data Analysis

To be able to reconstruct the sequence of human and natural processes that took place at the archaeological site, data analysis in archaeology entails the classification of the excavated or recovered archaeological record and the spatio-temporal sequencing of this record into archaeological assemblages. Consequently, archaeologists perform a site formation analysis, by constructing and placing different archaeological assemblages in a spatial and sequential framework, through classification and contextual association, and by using theories and models that help explain the presence of these assemblages at the archaeological site. This is done, partly during the fieldwork, by means of the 3D site plans and within a GIS

[10] If data and information obtained from individual sites are utilised to improve our background knowledge (see abstraction), it is also necessary to documented all sites routinely and consistently in the same manner, using digital methods and techniques.

application. This links the spatial locations of the mapped archaeological record to attribute tables containing quantitative and qualitative data and information on this record and the individual pieces of evidence that comprises this record.

It is believed that classification of the forensic record and the sequencing of this record into forensic assemblages will help to support or exclude hypotheses, identify people present at the crime scene during the criminal act and offer preliminary findings regarding the manner and cause of death. This will, therefore, establish logical and defensible (preliminary) scenarios concerning what (could have) happened at the crime scene. Forensic assemblages can be constructed and analysed in a similar manner as archaeological assemblages, by using a site plan and by assessing the relevant background knowledge. Forensic archaeologists will thus be able to build crime scene-specific site formation models and possibly sequence the physical evidence to activities that took place before, during and after the crime was committed. This could be done by asking oneself how a grave was dug, where the spoil from the grave was deposited, how the body was deposited in the grave, what type of tools was used for the digging, what taphonomic alteration modified the crime scene, etc.? Since it enables reasoning retrospectively, from the localised forensic assemblages to the originally deposited forensic assemblages, thinking in formation processes and forensic assemblages will help to understand how physical evidence was originally deposited by accounting for site formation processes and taphonomic transformations. In criminalistic terms, the use of forensic assemblages will help to access interpretations at the activity level. However, to be able to do this, a forensic archaeologist should also estimate the probabilities of transfer, persistence and recovery of the individual materials and traces that combine an assemblage. This is not necessary if the physical evidence is evaluated at the source level only. For instance, the presence of a thin polyester sewing thread around a skeletonised lower leg buried in a pit affected by scavenging could offer moderate support to the hypothesis that the buried leg was originally wearing a sock made of organic fibres and that, due to taphonomic alterations, this sock degraded leaving only the more durable polyester fibre. Additionally, the presence of insect pupae in an area affected by the scavenging gives – in view of the presence of the polyester fibre and the skeletonised leg – much more support to the hypothesis that these pupae are related to the scavenging than to the hypothesis that they are related to the original interment of the leg in the pit. In other words, by defining the pit and associated physical evidence as a forensic assemblage and by utilising taphonomic, degradation and ecological MR-theories, it is possible to classify and sequence (some of the) activities that took place in and near the pit.

1.3.6 Interpretation

As previously mentioned, the forensic record, as retrieved during crime scene investigation, will be a selection of the initially deposited physical materials and traces. This is not only due to the site formation processes and taphonomic transformations but also the investigative theoretical and methodological framework used during the crime scene investigation. It is often believed, especially by the scene of crime officers (SoCOs), that analysis and interpretation of physical evidence can be postponed until after the completion of the crime scene investigation. This should be done, according to the same officers, by collecting physical evidence at a crime scene as completely and systematically as possible, mostly by following hypothetico-deductive method (HD method). In addition, it is believed that the issues of (inductive) analysis and interpretation concerning the crime under investigation can be postponed until the results from the forensic laboratory arrive. This line of reasoning is, from archaeological point of view, incorrect. For example, the HD method, or the scientific method, was developed to test scientific hypotheses by observable data in scientific disciplines such as biology, physics and chemistry. This method works well in experimental sciences, under laboratory conditions. However, this is not the preferred method to reconstruct past human activities from (a partially degraded) forensic record or to help evaluate forensic evidence at the activity level. To reiterate, analysis and interpretation of the material record should start at the crime scene or place of incident and include a theoretical framework (e.g. MR-theories) to link the obtained forensic record to inferences about the human activities and natural processes that could have been responsible for the formation and alteration of this record in the first place. To help understand and analyse spatial and temporal patterns in the forensic record, a GIS application should ideally be used. GIS-based analyses could visualise the spatial layout of the recovered assemblages, help to clarify whether data could have been missed due to site formation processes or taphonomic transformations and help to decide whether an interpretation of the recovered record is plausible. In this way it would be possible to test inductive and deductive hypotheses and to assign a preliminary evidentiary value to the recovered forensic assemblage, given the hypotheses used. Hypotheses that survive falsification could be used to form scenarios about what could have happened at the crime scene, although it should also be kept in mind that excluding one hypothesis does not demonstrate the other to be true when there are more than two hypotheses, as it is often the case.

To summarise, the proposed interpretative framework will be able to (1) offer an intellectual tool for understanding the past (assessment), (2) detect assemblages in the forensic record against which evidential claims can be evaluated (analysis and interpretation) and (3) offer frames of reference against which observations of the criminal record may be contrasted to gain understanding (reconstruction). Of importance is that hypothesis testing at crime scenes will also expose ambiguities. This is essential since it will force the forensic archaeologist to consider alternative hypotheses or will point him/her towards new hypotheses (and therefore new scenarios) and consequently initiate a search for alternative physical evidence. In other words to update the current investigative methodology with new data or information and – if feasible – to modify this methodology, there should be a feedback loop between interpretation and final desktop analysis.

1.3.7 Integration

In archaeology, all individual lines of research, for example, physical anthropology, zooarchaeology, archaeobotany and ceramic analysis, will eventually be combined in one overall synthesis or excavation/recovery report. Most of the time finalising this report will be the responsibility of the archaeologist who was responsible for the excavation or recovery. Integration of data and information, as obtained during a forensic archaeological investigation of a police case, is currently not considered part of forensic archaeology. However, a forensic archaeological analysis and interpretation is sometimes combined with analyses and interpretations obtained from forensic anthropology (e.g. taphonomic analysis) and forensic ecology (e.g. post mortem interval (PMI) estimation), to offer a multidisciplinary evaluation of the encountered evidence and the most likely reconstruction. This constitutes a method of working that clearly benefits forensic investigation and the subsequent court proceedings.

Given its interdisciplinary framework, the archaeological process could be helpful in combining the separately carried out forensic evaluations – at the crime scene and in the laboratory – into one or more crime scene reconstruction(s). Such a reconstruction could best be performed within a Bayesian investigative framework, for example, using Bayesian networks (e.g. Dawid and Evett 1997) and GIS-related spatial statistics. An integration of all observations, hypotheses and background knowledge will surely offer a better reconstruction than the separate areas of forensic expertise can on their own.

1.3.8 Comparison

An archaeological report will compare the results obtained from the finalised excavation/recovery with results acquired from past excavations/recoveries that took place in the vicinity of the studied archaeological site. This is important since identification of (dis)similarities with previous archaeological studies can aid in further testing of hypotheses, offer new research questions and help the evaluation of the archaeological research strategy and field methodology.

Case comparison is also useful during a forensic archaeological investigation. It is relatively easy to accomplish if a GIS-based documentation system was used during the investigation. The use of GIS will not only allow a fast and reliable analysis of the investigated crime scene; it will also provide a fast comparison with previously investigated crime scenes by means of spatial statistics. In this way information obtained from different sites can be compared and combined to investigate cases at a supra-site level. For example, to state how common a tool mark encountered on the wall of the investigated pit is or to state if the dimensions of the burial pit deviate from the mean. Comparison with previous casework could be also relevant to formulate new forensic hypotheses, to see if possible physical evidence could have been missed and to evaluate the investigative methodology used during the investigated case at hand. Additionally, case comparison will also aid in an intelligence-led framework of policing and provide additional crime intelligence (e.g. Ribaux et al. 2010a, b).

Finally, after comparison, a (peer reviewed) forensic archaeological report will be ready to be submitted to the police or the court, depending on the jurisdiction.

1.3.9 Abstraction

Abstraction is the last step in the archaeological research cycle. By formulating new research ideas and hypotheses about the archaeological past, by updating the existing background knowledge and by helping to adjust future excavation and recovery strategies, this step offers the much needed feedback for future preliminary desktop research. Most of the time archaeological abstraction takes place by means of database updating, scientific publications, workshops and symposia.

Abstraction is also useful in forensic archaeology. New investigations will, undoubtedly, expose physical evidence finding the current theories and models deficient. Furthermore, it will now be possible to formally update the existing background knowledge to help formulate new research questions and forensic hypotheses and to help adjust and therefore improve our investigative methodology. Such abstractions must fall within the ethical and legal possibilities and limitations, for example, rules concerning the storing of personal data. Thus, forensic archaeological investigation should not be an end in itself. It should also be applied to supra-case level and used to generate new theories and models and to improve their predictive value.

1.4 Development and Innovation

To strengthen its scientific base, forensic archaeological research should also focus on the refinement of theories, models, methodologies and methods from both the natural and social sciences. To improve its operational field methodology, forensic archaeology, as a science, should focus on the pedological, ecological, biological and criminalistic theories and models that are needed to understand site formation processes and taphonomic transformations. New theories and models to predict where certain types of physical evidence may be encountered, given the type of the crime, and to assign (preliminary) evidential value to the obtained forensic record should be developed. Such knowledge would not only aid the development of scientific MR-theories but could also help explain the presence or absence of certain types of material evidence within the retrieved forensic record. Additionally, forensic archaeologists should keep up to date with new research and development within 3D documentation, predictive modelling, GIS science, geophysics and remote sensing and incorporate these innovations into their investigative framework. By integrating criminalistic and criminological theories and models into its interpretative framework and by using the archaeological research cycle, forensic archaeology, as a social science, should try to interpret and reconstruct human activities that took place at the crime scene or place of incident. The to be developed criminological MR-theories could be built by studying the location and distribution of relevant forensic assemblages resulting from different types of criminal activities. Depending on the questions to be addressed and the forensic hypotheses to be tested, such theories could be inductive (linking the empirical reconstructions to higher schemes or theories) or deductive (creating testable forensic hypotheses derived from higher schemes or theories). Of importance here is that forensic archaeology, in contrast to main stream archaeology, has direct access to information about past human behaviour (specifically information regarding criminal behaviour in the form of court testimonies, perpetrator confessions and trial rulings) and can therefore evaluate our interpretations and reconstructions of the past, something that is not possible in mainstream archaeology.

Four primary types of research could provide background knowledge, as described above: (1) case review studies concerning evidence patterns, evidence quantities, assigned evidentiary value, etc., as encountered at different types of crime scenes during past casework, (2) criminalistic studies concerning evidence patterns based on actualistic experiments using simulated crime scenes, (3) criminological studies concerning (offender's and witness's) testimonies and behaviours that could help explain the evidence patterns encountered at crime scenes during regular casework and (4) review of jurisprudence concerning the use of hypotheses and scenarios, as addressed by the court.

Of importance is here that this knowledge should be stored in indexed documents (by tagging or associating relevant information) and in quantitative and qualitative databases and that it should be linked to geographical and spatial locations, preferably within a GIS application. Such indexed documents and databases could be used as predictive tools during the different forensic archaeological investigative phases. During the preliminary desktop research, they could, for example, be used to predict where clandestine burials could be located and what the taphonomic condition of such sites might be. During the field investigation, the GIS-related databases could be used to map and document the encountered forensic record, to provide up-to-date view of this record and to test hypotheses. Post-excavation indexed documents and databases could be used as analytical tools, to analyse and reconstruct the obtained forensic record and to test additional forensic hypotheses. To ensure relevancy all databases, if ever developed, should constantly be updated with new data and information from new forensic archaeological investigations.

1.5 Conclusion

Forensic archaeology can be redefined as an archaeological discipline that combines the archaeological framework with the theory and methodology of criminalistics and criminology, all this within the context of (criminal) law. The aims of forensic archaeology should not only be to preform investigations in casework but also to conduct research. Forensic archaeology can only evolve scientifically and methodologically if it is broadened beyond the general boundaries of a technical investigation. If forensic archaeological practice is redefined as a cyclic scientific process, it will not only be able to generate new forensic knowledge and new ways to look at the physical evidence, but it will also be able to improve future casework by means of case abstraction.

The introduction of the archaeological research cycle, crime scene formation models, taphonomic transformations and the concept of background knowledge and forensic assemblages in forensic archaeology will without a doubt aid in the location of physical evidence that is capable of discriminating between forensically relevant inductive and deductive hypotheses. Such an approach is strengthened by linking the obtained forensic record to human activities that took place before, during and after the crime was committed. To help consider where to look for certain types of physical evidence and what its evidentiary value, given the forensic hypotheses addressed, could be and to evaluate why certain types of physical evidence were not discovered, the proposed framework should be integrated with the Bayesian paradigm.

Acknowledgements The author would like to express his gratitude to Coen Nienaber (Netherlands Forensic Institute) for his helpful and constructive comments on the earlier version of this chapter that contributed to its final version.

Part I
Forensic Archaeology and NDTs

Chapter 2
Forensic Archaeological Remote Sensing and Geospatial Analysis

Alastair Vannan

Abstract Forensic archaeological geospatial analysis can contribute significantly to searches for people who are missing, presumed murdered, and investigations for other buried or concealed items. This sub-specialism of archaeology is an extremely wide-ranging and multi-faceted sphere of study and practice that can incorporate numerous techniques and types of data, and as a result, individual aspects of this approach are often treated in isolation. This can present a fragmented impression that implies that each remote sensing technique or spatial analytical approach can stand alone as a self-contained system. This chapter discusses how such techniques are most effective when combined as a suite of approaches and applied within a fully integrated system. Suggestions for how we might begin to establish guidelines for best practice at a time of conspicuous growth in technologies and data types are also discussed. This provides a context for the summarised examination of two specific approaches within forensic archaeological geospatial analysis: firstly, the capture and analysis of LiDAR data and secondly, the use of red, amber, green (RAG) mapping. LiDAR data is ubiquitous in many conventional archaeological studies, but the capacity to capture this data using unmanned aerial vehicles presents us with the potential to produce data of a quality that is suitable for forensic archaeological purposes. Red, amber, and green (RAG) classifications of soil diggability are used as a search tool in forensic geology. This chapter suggests moving away entirely from the use of RAG mapping as a tool to define prioritised search areas and towards its use as part of human-centred landscape analyses once prioritised search areas have been defined.

Keywords Forensic archaeology · Forensic remote sensing · Geospatial analysis · Landscape archaeology · LiDAR · RAG mapping · Forensic search · Missing persons · GIS · Digital 3D modelling · Earth observation · Unmanned aerial vehicles · UAVs · Drones

---

A. Vannan (✉)
Cellmark Forensic Services, Chorley, UK
e-mail: avannan@cellmark.co.uk

P. M. Barone, W. J. M. Groen (eds.), *Multidisciplinary Approaches to Forensic Archaeology*, Soil Forensics, https://doi.org/10.1007/978-3-319-94397-8_2

2.1 Introduction

The following text derives from papers presented to the European Meeting on Forensic Archaeology (EMFA) at Rome in August 2017 and comprises a discussion of the application of remote sensing data and digital geospatial analysis within forensic archaeology. These approaches enable sophisticated landscape analyses that can contribute to the search for people who are missing, presumed murdered, and also the search for buried or concealed items.

This is an extremely wide-ranging and multi-faceted sphere of study and practice, and as a result, individual aspects of this approach may sometimes be treated in isolation as independent studies. The necessarily restricted scope of a research project may seek to provide detailed understandings of, for example, the capabilities and limitations of one specific type of remote sensing data. Such detailed studies are essential to enable us to identify the most appropriate and effective techniques to use within the specific parameters of each forensic investigation. To this end, the author is engaged in ongoing studies into the effectiveness of numerous types of aerial-captured remote sensing data in identifying a wide range of ground-level and sub-surface targets.

The individual treatment of different aspects of geospatial analysis has the potential, however, to present a fragmented view of these techniques and approaches and to provide the implicit impression that a certain remote sensing technique or spatial analytical approach can stand alone to be applied as a complete system. Each data type is subject to inherent limitations and, where possible, should not be used in isolation from other data that allow comparative analyses to be undertaken. Equally, the data should never be used in the absence of interpretation by an analyst with the correct level of experience in landscape analysis and forensic science. The latter point is of particular importance in the current age of rapidly developing modes of data capture, processing, and visualisation.

Geospatial data types and techniques are expanding rapidly, and this is, therefore, an important time to provide an overview of some key issues affecting the use of these tools. This may assist in generating dialogue to determine good practice in the area of forensic archaeological geospatial analysis and to identify and pre-empt some potential pitfalls. To this end, this discursive and generalised chapter will be presented in two parts. The first part will aim to provide a discussion of overarching considerations for the practical application of forensic remote sensing and geospatial analysis. This will provide a context for the second part, which will examine the use of LiDAR data (sometimes expanded as light detection and ranging or laser imaging detection and ranging) in the search for clandestine graves and the use of red, amber, green (RAG) mapping to present and utilise geological data in search investigations. The text is based primarily on direct experiences in forensic casework in the UK and with brief reference to ongoing research.

2.2 The Growth of Remote Sensing Data and Geospatial Analytics

In May 2016, the president of the Royal Geographical Society, Nicholas Crane, presented the keynote speech at the Esri UK Annual Conference in London. He asserted his conviction that a new age of geography was emerging globally, with vast and increasing means of earth observation, mapping, and analytics developing exponentially. Indeed, the growth of geospatial computer science is conspicuous. Locational information is ubiquitous in online popular culture and commerce from map-based applications, displaying the distributions of businesses and services, to the management of personal digital photographs by the place of capture, targeted advertising based on location, and increasingly sophisticated navigation apps. Remote sensing data capture and analysis is already controlling driverless vehicles on public roads (e.g. systems by Tesla and Velodyne), and the scientific and commercial use of remote sensing data spans almost every area of engineering, land management, natural resource management and exploitation, natural disaster monitoring and response, and military applications. As a result, geospatial computer science provides a potential career path for huge numbers of technologically literate individuals with skills in programming, software engineering, and data science/data analytics where statistical techniques and machine learning are applied to large datasets.

This provides a wealth of emerging talent to aid in the development of new technologies for forensic searches, but it is vital that this talent is harnessed and married with expertise in the disciplines of forensic geoscience, i.e. the application of geology or archaeology, if reliable results are to be produced. The hardware (e.g. unmanned aerial vehicles (UAVs), cameras and other sensors), data (e.g. photography, multi- and hyperspectral imagery, LiDAR data), software development (geographic information system (GIS) and 3D modelling applications), and new data processing techniques are essential to ensuring that we are operating at the limits of current technological capability, but these do not supersede the role of the spatial analyst. Effective interpretation of results is not possible in the absence of expertise in the natural and anthropogenic formation processes responsible for the vast array of ground disturbance and landscape change detectable through the use of remote sensing data. With considerable levels of programming and the provision of data relating to representative anomalies, an algorithm may be able to assist in detecting patterns indicative of ground disturbance (see e.g. Paz et al. 2010; Zakšek et al. 2011; Borgys et al. 2012; Doneus 2013; Guo et al. 2016). However, it cannot differentiate effectively between the many potential causes of the disturbance, interpret the specific activity responsible, or identify likely relative chronological sequences for a series of features.

Archaeologists are able to categorise a vast range of features as belonging to a recognisable 'type', but these types are abstractions, rather than being tightly defined according to immutable physical characteristics. Concessions can be given that allow an ill-fitting feature to be accepted into a category through an understanding of wider landscape changes and, therefore, anticipated changes to the expected form. For example, a field boundary bank might be expected to comprise a raised earthwork with a conical profile and stone retaining walls, be topped with a hedge, feature a parallel ditch, and delineate a parcel of land. However, if ploughed out, the same type of feature may be recognisable as a low linear rise with a corresponding parallel linear hollow, no stone walling or hedge, but perhaps one or two mature trees representing the remnants of the overgrown hedge. We can then extrapolate from the level of damage that a subsequent episode of human activity has occurred, i.e. cultivation.

The development of sophisticated modes of machine learning, utilising artificial neural networks to recognise complex patterns in imagery, has the potential to aid in forensic remote sensing data analysis. However, such technologies tend to be developed for major commercial applications, and the sphere of forensic investigation is unlikely to be the recipient of the enormous funding necessary for such development. Even if such approaches were developed, the considerable variability in landscape morphology, particularly at the very detailed topographic scales required for forensic archaeological investigation, means that it would be exceptionally difficult to produce an automated system that could be applied effectively. This would be particularly challenging given the wide variety of different landscape types and the presence of anomalies representing multiple overlying and intercutting fragments of features of varying date and resulting from disparate activities.

Conversely, expertise in digital geospatial analysis is not common to all practising archaeologists or forensic archaeologists. The possession of archaeological experience is not enough to enable valid geospatial analysis in the absence of familiarity with the software and data types being used or familiarity with the principles of landscape archaeology.

2.3 Forensic Archaeological Landscape Analysis and Geospatial Capabilities

Desk-based assessments are a ubiquitous part of any archaeological or geological investigation, and this applies equally to forensic archaeological investigations (Hunter and Cox 2005, 33–36; Hunter and Martin 2002, 89–95; Litherland et al. 2012, 30–31). The scope of these investigations can be variable, and one of the significant differences between forensic and commercial archaeology in the UK is the requirement for extremely rapid deployment to aid in the investigation of a crime scene, for example, if the presence of human remains or a suspected clandestine grave has been identified. This can mean the need to respond immediately to a call from the police with little time to undertake any detailed desk-based analysis. In this instance, it may be possible only to undertake a rapid check of modern and historical maps to gain an understanding of the general landscape development that will inform the interpretation of the physical features that will be encountered. However, when the forensic archaeologist is called upon to assist in wider search investigations and a precise target area has not yet been identified, this will generally provide more opportunity for pre-planning and to undertake a thorough desk-based assessment.

Desk-based assessments in all areas of archaeology typically include as standard the examination of modern and historical mapping, modern and historical aerial photography, geology and soil maps, records of known sites of archaeological interest (held within Historic Environment Record (HER) databases in the UK), and varying levels of supporting historical information depending upon the nature and location of the area of interest. Where available, other data might also be utilised, such as LiDAR data that has been previously captured by the UK Environment Agency for the purposes of flood modelling. In some instances, detailed landscape analyses will include the capture of bespoke remote sensing data, such as photogrammetry data captured by UAV, although this would normally comprise a specific phase of targeted survey of identified sites, rather than informing a generalised landscape analysis. Some large landscape research projects are exceptions to this, such as the Stonehenge Hidden Landscape Projects (Gaffney et al. 2012), which utilised multiple geophysical techniques and laser scanning to investigate an area of approximately 8.2 km^2.

Commercial (predevelopment) archaeology has developed at a fast pace in the UK, adopting equipment and techniques for survey and earth observation from a wide range of disciplines including geography, engineering, and environmental and agricultural monitoring. Academic research has also experimented with, and developed, these new approaches, undertaken comparisons between techniques, and undertaken trials to help establish the capabilities and limitations of data, modes of capture, and processing (see e.g. Rowlands and Sarris 2006; Doneus 2013; Bennett 2011; Historic England 2017, 16–20). The development of new capabilities has enabled the character of desk-based assessments to develop far beyond what was possible as little as 10 years ago and to enable considerably more sophisticated levels of remote landscape and spatial analysis.

In the UK, forensic archaeology has not, historically, kept pace with the advances in geospatial technology that are a feature of specialism for some commercial and academic archaeologists. The reasons for this appear simple:

- There are very few forensic archaeologists with dedicated specialisms in landscape analysis.
- Forensic archaeologists are not necessarily able to invest in geospatial equipment, software, and training to the same level as large archaeology units/academic departments.

There are very few practising forensic archaeologists in comparison with conventional archaeologists, and therefore, we cannot expect as diverse a range of archaeological specialisms within forensic archaeology. Where forensic archaeologists develop dual specialisms, these, understandably, tend to relate to anthropology and taphonomic processes. Forensic archaeology also tends not to be as well funded as development-led and some research archaeology. Teams tend to be small or comprise sole practitioners or individuals within the police who work primarily in other roles. Such teams or individuals cannot justify the investment in sophisticated survey instruments and software that large companies employing sometimes hundreds of archaeologists are capable of.

This means that forensic archaeology in the UK has needed to catch up with other areas of archaeology, and we might find that aspects of GIS or airborne remote sensing are used but sporadically and not necessarily applied as standard within a holistic system that provides an integrated and developed approach to landscape analysis. We might see the use of GIS and airborne remote sensing data in forensic archaeological research or practice, but we do not know to what extent the potential of these instruments, software, data, and information derivable from the data is being utilised.

2.4 The Development of Best Practice Within Forensic Archaeological Geospatial Analysis

GIS and other software for viewing or interrogating geospatial data have many hundreds or even thousands of tools, and the software can be used to extremely good effect or in relatively simplistic, or even naive and misleading, ways. For example, the effective post-processing of height data, such as LiDAR or photogrammetric data, can make the difference between the provision of highly informative data and what are little more than pictures of limited analytical value. Additionally, a lack of understanding in interpreting the formation processes responsible for the features visible in the data will render well-processed data meaningless. In order to produce high-quality results, the integrity of the data, subsequent processing, and interpretation at every stage must be demonstrably effective and of a rigorously high standard.

The following areas will be important to consider in any attempt to develop consistency and integrity in the practical application of forensic geospatial analysis.

2.4.1 The Capability of the Datatype to Achieve the Desired End

Aerial-captured remote sensing data must be of sufficient resolution to enable the identification of the range of sizes and types of feature expected given the details of the investigation. The resolution refers to the pixel size provided by the data. Features will only be identifiable within any dataset if they are above a certain size, and this size will vary according to the resolution of the data. For example, aerial photographs with a resolution of 50 cm will comprise pixels with sides measuring 50 cm. This means that a grave measuring 1 m long and 0.5 m wide would be represented by a maximum of two pixels. It is extremely unlikely that this resolution will be sufficient to allow such a feature to be identified by examining either the photographic image or any associated height data.

The data selected must also be appropriate to the ground conditions. If, for example, LiDAR is used and the woodland canopy is extremely thick, such as that presented by a mature plantation woodland of closely spaced conifers or a dense woodland of young trees in full leaf during the summer, then this may not provide data of sufficient quality to enable meaningful analysis. Similarly, a photogrammetry survey in an area of high vegetation will not produce reliable height data for the ground surface. These are stark examples, but many more subtle considerations of ground conditions should be undertaken in advance of any data capture or analysis, and a range of these is the subject of ongoing research (Quick 2017; Dr. D. Jordan, personal communication).

2.4.2 Processing the Data and Resultant Level of Accuracy

Most types of aerial-captured remote sensing data will require some level of processing in order to produce an output that is able to be viewed and interrogated in a GIS or using a 3D model viewer. This may include locating the data to known survey control points in order to produce scalable data and/or data that is tied to a geographic coordinate system. Photographic data will need to be processed in order to enable height data to be generated from a photogrammetric survey, and LiDAR data will require processing in order to ensure the cleanest data possible is then taken forward for analysis, with a minimum of distortion and a consideration of inconsistency between the multiple sensors of the instrument and/or between data captured in overlapping flight paths. The way in which these, and other issues, are addressed during initial processing will have an impact on the clarity, accuracy, and reliability of the data and, therefore, every subsequent interrogation of that data. Establishing the margin for error and the limitations of the data is extremely important to inform the appropriate use of the data.

2.4.3 Post-processing and Visualising the Data

Every type of aerial-captured remote sensing data will be able to be processed within a GIS or 3D model viewer to produce different visualisations. Each visualisation will allow a different perspective of the data, enable different aspects of the data to be focused upon and, therefore, enable more nuanced and developed interpretations of the data (see Figs. 2.1, 2.2, 2.3, and 2.4 for multiple visualisations of photogrammetry data). Conversely, each individual visualisation is limited in the information that it can provide. Therefore, if only one visualisation of the data is used, there is a risk that this may exclude important indicators and that the analysis

Fig. 2.1 A linear feature (possible cut feature) is visible aligned north-west/south-east in RGB photography

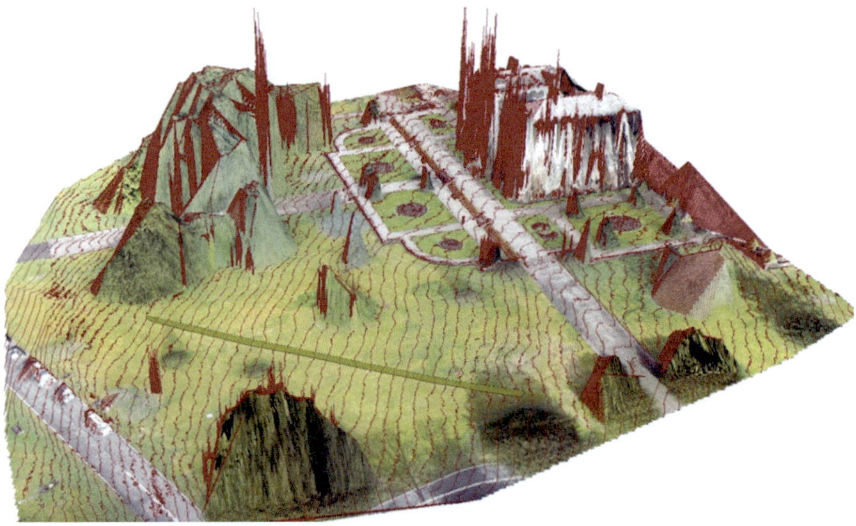

Fig. 2.2 Viewing the linear feature against detailed contours (10 mm intervals) derived from the same data reveals that it runs in the direction of a slight slope. This would be consistent with a drainage channel

will be poorly informed. A stark example of this is the use of jpg versions of LiDAR data. These are static images that cannot be interrogated adequately and should never be used in isolation in any serious analysis. Even rasterised LiDAR point data is subject to the inherent limitations at the point of processing (e.g. resolution) and ideally the point cloud, e.g. in LAS data format, should be used by the analyst

Fig. 2.3 Examination of an elevation model derived from photogrammetry data shows subtle indications of ridge and furrow running through the extent of the linear feature. This demonstrates that the linear feature does not cut through the ridge and furrow

in the production of multiple visualisations, rather than a sole reliance upon derivative datasets produced by a third party.

2.4.4 *Comparative Analysis*

The use of multiple datasets to examine a single area enables comparative analysis to be undertaken that can provide enhanced understandings of the search area. This can provide the opportunity to test hypotheses derived from the analysis of one dataset against another set of data in order to provide a nuanced and more highly developed understanding of the feature and, therefore, to enable increased confidence in the final interpretation. Where only one type of data is available, it is often possible to utilise multiple visualisations of the raw data to undertake comparative analysis.

2.4.5 *Use of Controls*

A site visit prior to undertaking an analysis of remote sensing data can be highly beneficial. However, this will not always be possible in a forensic context due to operational necessities for covert observation. If a visit is undertaken, then it can be beneficial to identify control features that may have similar characteristics to significant features, such as clandestine burials. These will not be relied upon too heavily

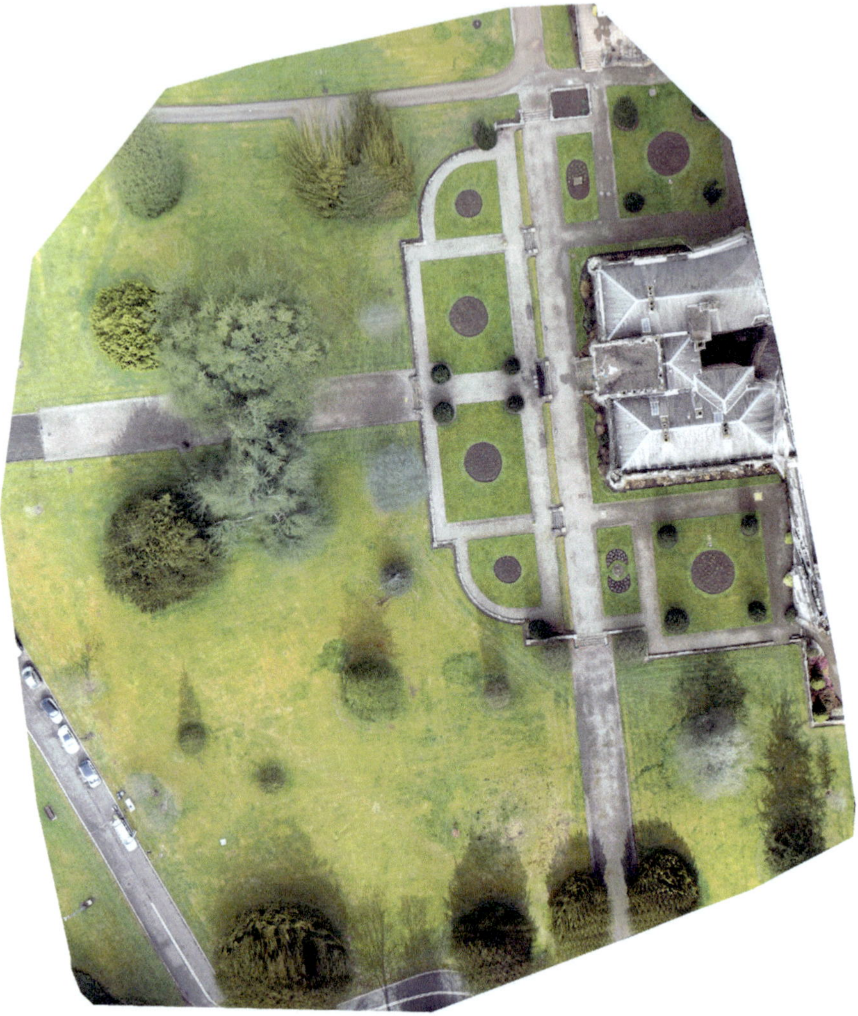

Fig. 2.4 The linear feature is revealed to be the result of repeated footfall across an informal routeway (shortcut), which was not recorded on any mapping

but can give an understanding of, for example, how a shallow hollow will appear in this particular dataset in this particular landscape. It is also useful to establish how a patch of differential vegetation growth within the local environment will appear in the data set. Measurements and photographs of the control feature and a way to mark it clearly in the dataset, such as placement of a highly reflective survey target adjacent to the feature during the capture of data, will enable effective identification and examination during analysis. This may assist in determining the most effective form of processing and visualising the data for the highest degree of clarity of potentially significant features.

2.4.6 Competence to Interpret Data

The competence of the data analyst may seem like an obvious consideration; however, technological approaches can run the risk of becoming dominated by the hardware and software, and the risk exists that the competence of the data analyst can become overlooked. Landscape analysis and interpretation is a discrete area of specialism within archaeology. As with every other area of specialism, considerable time is required to develop relevant skills and experience. Thousands of different features that have been produced by a myriad of formation processes exist in the massively varied landscapes of the UK. Without a thorough understanding of the range of features that is likely to be encountered within, for example, upland moorlands, lowland agricultural landscapes, residential environments, industrial landscapes, coastal zones, woodlands, and wetlands, the analyst will not be capable of providing any meaningful interpretation of the data. This means the ability to distinguish between natural and anthropogenic formation processes, differentiate differing anthropogenic activities, identify likely natural and anthropogenic disturbance and transformation of the features, and establish relative chronologies for all of these identified processes.

Significant time spent examining thousands of features from widely differing landscapes is required before an analyst can develop a high level of competence in this area. It is essential to contribute to a police investigation by cutting through the noise of irrelevant anomalies in the data to identify and prioritise those with the highest potential to represent features significant to the investigation. In the absence of adequate interpretation, significant anomalies could be overlooked, and a subsequent lack of sufficient triage could lead to the police being presented with an unnecessarily large number of anomalies for ground-level investigation. The purpose of such remote sensing analysis should be to enable a targeted approach for ground-level searches informed by a clear and demonstrable rationale. Anything less will be conjecture resulting in essentially randomised targets for ground-level search.

2.4.7 Avoiding a Myopic Search Focus

A landscape archaeological approach should seek to understand the encountered environment as a whole and should not adopt a myopic focus on a search for the grave site alone. The archaeologist should seek to understand as much as possible about the human experience in the landscape, and this is what differentiates the archaeological approach from other methods. Of course, in a search for a person who is missing, presumed murdered, the deposition location will be the most significant target to identify. However, there may be other characteristics of the environment that aid in identifying areas of particular potential significance and in prioritising features of high potential. Examples of landscape considerations that could inform a study include gaining an understanding of:

- Sense of seclusion
- Formal and informal pedestrian routeways
- Vehicle access and parking opportunities
- Physical barriers to movement
- Landscape development prior to and since the time of the suspected offence
- Characteristics of the landscape that provide opportunities for easy concealment (hollows, ditches etc.)
- Statistical information to aid in prioritising portions of a search area
- Complimentary spatial data, e.g. results of comparative soil or botanical analysis, cell-site data, automatic number plate recognition, witness sightings, and intelligence

The identification of informal routeways, for example, can enable an understanding of the way in which people actually engage with the landscape, rather than the impression provided by depictions of formal routeways presented on mapping, which may not reveal an accurate, or complete, picture. This can be essential in tracing the routes used by suspects or offenders and also in identifying discarded evidence that could be otherwise overlooked.

2.4.8 Peer Review

The range of possible errors in data preparation and interpretation means that peer reviewing of processes and results is essential in GIS and remote sensing analyses. This can be difficult given the relative isolation of many forensic archaeologists and also the difficulty in identifying individuals holding the necessary combination of relevant experience. This can be alleviated by using multiple reviewers so that the processes can be checked by a person familiar with hardware, software, or data type, and the results can be checked by an individual familiar with the identification and interpretation of the physical evidence for varied landscape phenomena.

2.4.9 Data Integrity and Metadata

Various guidelines exist for the treatment of digital data, and versions of these guidelines should be applied in forensic remote sensing and geospatial analysis. One of the most important of these is that the source data should be retained untainted in its original format and copies should be made for any post-processing or manipulation. This ensures the integrity of the data should it be necessary to revisit it and to replicate subsequent processes and analyses.

Metadata is the provision of information within the data that details how that data has been created, i.e. the identification of the source data, the methods used to further process the data, any changes made to the data, the name of the analyst undertaking the modification or creation of data, the date of creation, etc. Metadata is already recognised as being an extremely important aspect of data production and modification (e.g. Walker 2012). However, when adhering to tight timescales, this can be an aspect of analysis that suffers neglect. If we are to enable subsequent analysts, such as defence specialists, to replicate what we have undertaken in terms of processing and visualisation, then the provision of accurate and comprehensive metadata is extremely important.

2.5 Forensic Archaeology and LiDAR Data

LiDAR, which is generally understood to be an acronym for light detection and ranging, is a form of remote sensing that uses pulses of infrared light from a laser to capture three-dimensional data points (Davis 2012, 4). LiDAR data has been used to contribute to archaeological projects for many years (e.g. Challis 2004, 2005, 2006; Bennett 2011, 17–21), and the author used LiDAR data as a part of standard landscape archaeological practice from at least as early as 2011 (e.g. Schofield and Vannan 2011). One of the key differences between conventional archaeological use of LiDAR data and forensic archaeological practice is the size/scale of the sites being targeted by prospection. LiDAR data is extremely effective in identifying remnants of landscape features of archaeological interest, such as building complexes, Roman roads, ridge and furrow, field boundaries, etc. Even if these features have been truncated by later activity, their scale can enable the general form to be extrapolated from the surviving visible portions. This means that LiDAR data that is widely available, such as that provided by the UK Environment Agency, with a resolution of between 25 cm and 2 m can be sufficient to contribute to landscape analysis and to inform our understanding of the historical development of a study area. Conversely, in most forensic searches for clandestine burials and associated indications of human activity, the features of significance are likely to be small in size and few in number. Therefore, the resolution of the LiDAR data in a forensic archaeological search will need to be sufficient to enable the confident identification of features measuring around 1.8 m long by 0.8 m wide and less. Data with a resolution between 25 cm and 2 m will not be sufficient to enable such identification as, even at 25 cm resolution, this would provide a maximum of 21 accurate points of height data for a feature of this size. If features do not sit significantly higher or lower than the surrounding landscape, then such topographic subtleties will be almost impossible to identify from such data.

As standard archaeological analysis pertains to premodern features, the date of the capture of the data is not generally of particular importance. Exceptions to this might be ensuring that LiDAR data captured for wooded areas has been undertaken at times of year with low leaf coverage or when an understanding of the landscape before or after a key event is required (e.g. assessing damage caused by a recent development). In forensic searches, however, the date of data capture is vitally important. Data captured prior to the date of the offence may allow a general familiarity of the landscape to be gained but will clearly not allow disturbances associated with the offence to be identified.

In order for LiDAR data to be of value to a forensic archaeological search, it must, therefore, be of sufficient resolution and quality and post-date the time of the offence. These qualities cannot be guaranteed from publicly available archived LiDAR data in the UK, and bespoke capture using UAVs offered the potential to provide data of sufficient quality. The Anthropology, Archaeology and Ecology Department at Cellmark Forensic Services, therefore, partnered with Flythru Heli-Drone Imagery to undertake trials to test the capabilities and limitations of LiDAR data captured by UAVs in the context of forensic searches. Detailed treatment of this research will be the subject of future papers, and a summarised description is presented here.

Three proxy graves were established within a woodland setting close to Barnsley, in South Yorkshire, England (Fig. 2.5). These features were created to mimic hollows and adjacent mounds associated with clandestine burials. Grave 1 measured 1.6 m by 0.6 m wide and was approximately 110 mm deep with an adjacent mound 220 mm high, and the hollow featured sharply cut vertical sides with an approximately 90° break of slope base. Grave 2 measured 1.8 m by 0.8 m wide and was also approximately 110 mm deep with an adjacent mound 220 mm high, but the profile of the hollow features very gradually sloping sides with breaks of slope at the top and base of the slope that were not sharply defined. This would, therefore, be more subtle and difficult to identify as an anomaly. Grave 3 was dug out to a depth of 0.6 m and was then completely backfilled. This created an area of general disturbance measuring 3 m long by 2.5 m wide containing a mound of soil measuring 1.5 m long by 1 m wide and 180 mm high. A fourth extremely small proxy grave

Fig. 2.5 Aerial view of woodland used for the LiDAR trial

was dug close to a mature tree and measured 0.78 m long and 0.6 m wide by 80 mm deep and with a small mound 100 mm high.

A LiDAR survey was undertaken in February 2016 using a UAV LidarPod manufactured by Routescene and carried by a bespoke UAV assembled by Ben Bishop from Flythru. The data was processed to produce a point cloud (a 3D model comprising a mass of individual discrete points; Fig. 2.6). This was processed to exclude trees and comprise only the points representing the ground surface. This bare earth model provided a point density of 78.3 points per square metre, which is extremely dense for LiDAR data captured over woodland. This enabled a rasterised (continuous sheets of data representing a surface without gaps rather than a series of points) visualisation of the data at approximately 113 mm (11.3 cm) resolution (Fig. 2.7). However, as the pattern of the points captured does not conform to a regular grid, the precise size of polygonal cells formed between adjacent points as a triangulated irregular network (TIN) will vary across the dataset. This comprised a considerably higher level of detail than LiDAR data ordinarily available commercially (Figs. 2.8 and 2.9), which was a critical requirement if subtle features such as the proxy graves would be able to be identified. Data at a resolution of 25 cm, which is high quality for data captured by aeroplane or helicopter, would have a point density of around 16 points per square metre. Subsequent data capture by UAV applied during UK casework produced an extremely high point density of 416 points per square metre, which equates to 49 mm (4.9 cm) resolution.

The data was exported as LAS files for incorporation into ArcGIS and was then post-processed to produce shaded relief DTMs. Numerous visualisations of the shaded relief models were produced utilising a range of colour ramps and elevation intervals to enable different portions of the height data to be highlighted. A variety of azimuths for the synthesised light source allowing projections of shadows to be cast across the digital model were also utilised to assist in the identification of subtle

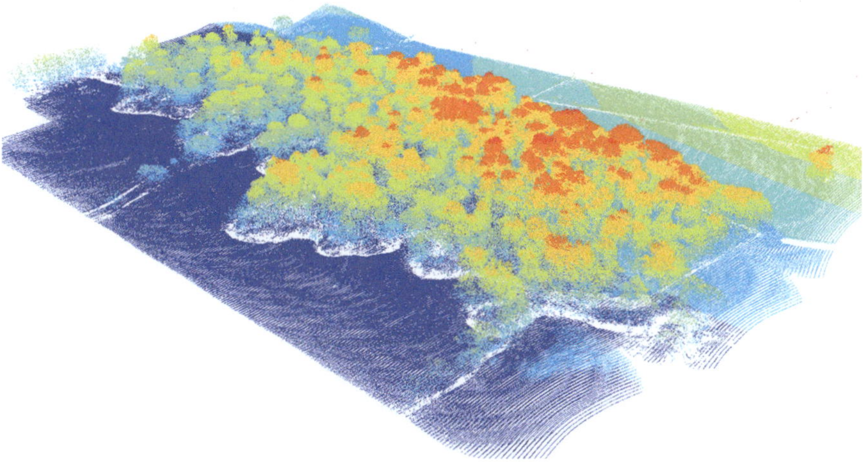

Fig. 2.6 Point cloud derived from the LiDAR survey of the woodland

Fig. 2.7 Digital terrain model (raster DTM) with trees having been stripped out of the data

Fig. 2.8 A visualisation of the LiDAR data at a resolution of 25 cm, which is consistent with high-quality commercially available data

topographic variations. The examination of the data demonstrated that graves 1–3 were visible as anomalies (Fig. 2.10) but grave 4 was not.

In addition to using the LiDAR data to generate topographic models, the intensity of the returned signal was also examined. This indicates the strength of the laser signal having struck a surface and returned to the instrument. The intensity of the returned signal will vary depending upon the characteristics of the surface struck, such as its level of reflectance (see Kashani et al. 2015), and this data can, therefore,

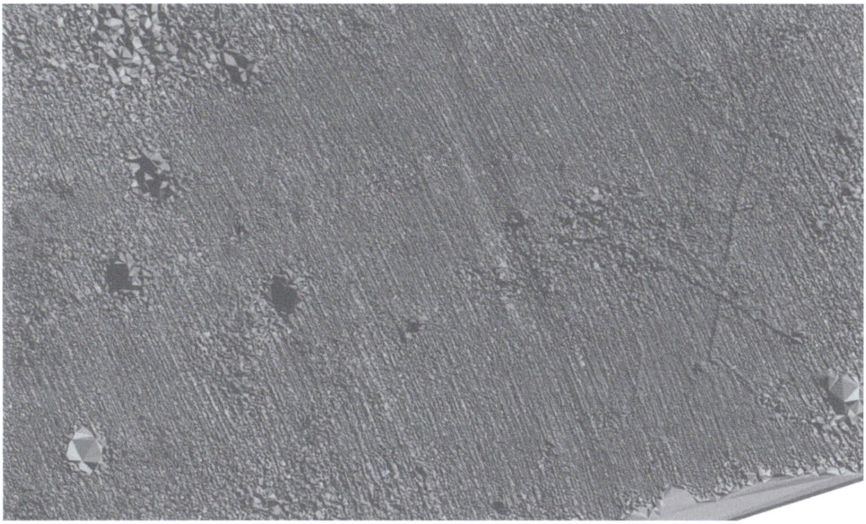

Fig. 2.9 Visualisation of the LiDAR data at 5 cm resolution, showing the stark increase in the level of detail with fallen trees and clusters of branches clearly visible

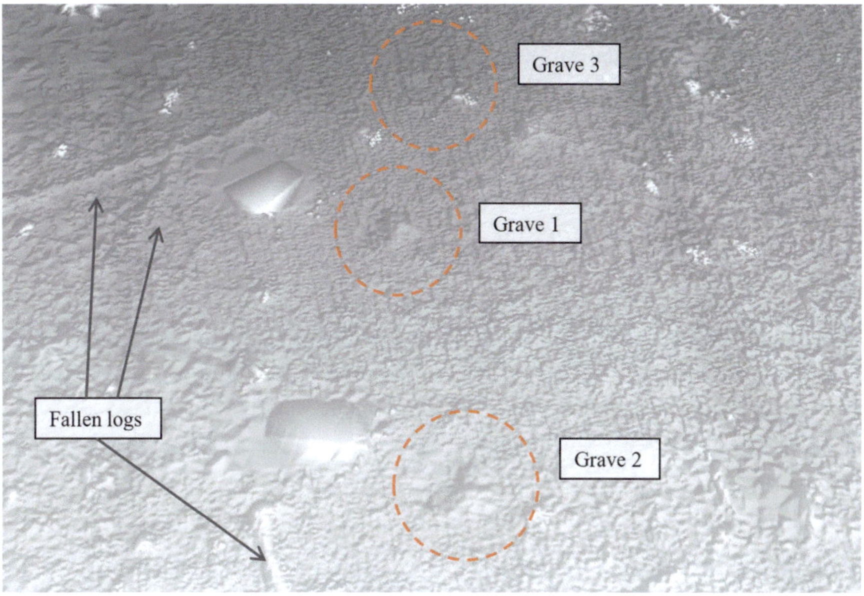

Fig. 2.10 Proxy graves 1, 2, and 3 visible in the DTM

indicate relative differences in the character and texture of the ground surface. This data was analysed as point cloud data and raster surfaces, and each of these data types was visualised using a variety of colour ramps and a range of intervals between the intensity values. This enable significantly enhanced views of the data to be

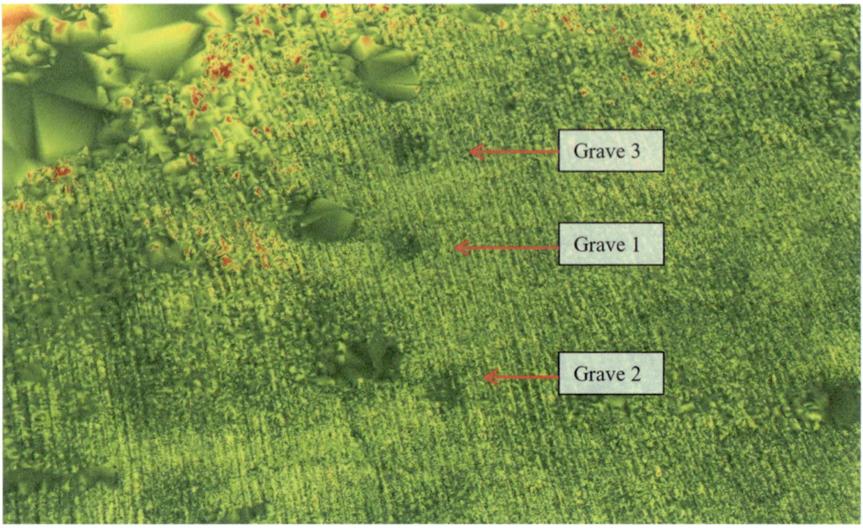

Fig. 2.11 Proxy graves 1, 2, and 3 visible in the intensity data

produced. The slight differences in the soil texture and character in the areas disturbed by the establishment of the proxy graves produced highly conspicuous anomalies in certain visualisations of the intensity data (Fig. 2.11). Further research will be required to examine the longevity of these detectable differences as settlement, compaction, and patterns of vegetation growth affect the disturbed areas.

2.5.1 Conclusion to LiDAR Discussion

This research was successful in demonstrating that features of the form and scale of the proxy graves were detectable within LiDAR data captured by UAV. This method of capture represented a significant improvement in the quality of LiDAR data as available publicly in the UK and provides a viable option for spatial analysis in forensic archaeological searches.

2.6 Forensic Archaeology and RAG Mapping

The use of red, amber, and green (RAG) classifications for soil diggability is an extremely useful tool when understanding the physical characteristics of the landscape and the potential influence that these characteristics may have on the ability of an offender to dig a clandestine grave. This system applies red, amber, and green colour coding to indicate areas of easy, intermediate, and difficult digging (Donnelly and Harrison 2013). As forensic archaeologists, we have the opportunity to contribute to the application of this geological approach by making adaptations deriving from theory and practice from our own discipline. This combination of forensic geology and archaeology has the potential to draw nuanced information from the landscape that could be of significant use to police investigations for people who are missing presumed murdered.

The key aspect of this suggested approach is the application of geological character as one element of a human-centred landscape analysis to inform and direct the character of ground-based searches. It is the human-centred approach that distinguishes the archaeological application of RAG mapping from the geological application. It is the author's assertion that geological character, in terms of diggability, should not influence the process of search prioritisation. The archaeological approach should instead apply diggability subsequent to the definition of prioritised search areas.

The prioritisation of search areas should integrate as many different stands of information as possible in order to identify areas with the highest potential to contain human remains, including, but not restricted to, intelligence, cell-site and automatic number plate recognition (ANPR) data, offender profiling, statistical information (such as the CATCHEM database in the UK), reconstructions of the historical landscape, and comparative geology, soil, and vegetation analyses.

If we define search areas for concealed remains based on diggability, then we are presuming that human behaviour is irresistibly influenced by the physical environment. Such ideas of environmental determinism were largely set aside in archaeology with the development of post-processual theories in the late 1970s and 1980s as it began to be accepted that human societies and behaviours are not determined by external influences alone (see, e.g. Johnson 2002, 98–115). It is, therefore, essential for the forensic archaeologist to put human activity and behaviour at the centre of the consideration of deposition sites, rather than assuming that offender behaviour is determined purely by the environment (Fig. 2.12).

There are some rare instances in which we might expect the sub-surface ground character to affect the choice of deposition site, for example, if the suspect is a geologist, archaeologist, or other geoscience specialist. Similarly, if the suspect has exceptional awareness of local ground conditions, such as a countryside ranger or groundskeeper, then the sub-surface conditions may be significant. Without such specific prior understandings of ground conditions, it is extremely unlikely that the choice of deposition site will be informed by the geological character. Instead, some of the primary concerns for the offender, particularly if the murder is not premeditated, might be:

- Is the deposition site accessible?
- Can a vehicle be parked nearby?
- Is the site familiar?
- Is the site secluded?
- How far will to body need to be carried?
- Will the offender be observed before, during, or after deposition?

Fig. 2.12 This map shows that the woodland in the centre of the image features geology of variable diggability. This would not be known to an offender, and other factors, such as accessibility, might play a primary role in determining the general deposition area

If the local geological character has not influenced the initial choice of deposition site, then the offender may only become aware that it is difficult to dig at the point of deposition. In this situation, it may be unlikely that an offender will restart the hazardous process of identifying and moving the body to a completely separate deposition site, with no certainty of encountering better ground conditions. This situation presents the potential for particular activities at the point of deposition that may produce identifiable material remains that will be important to consider during an archaeological search.

Evidence of aborted dig sites may be present, where attempts have been made to dig a grave but been frustrated by the presence of hard geological deposits or roots, etc. The inability to dig successfully could lead to the use of pre-existing hollows or soft spots in the landscape for deposition. These may be of natural origin, such as tree throws and stream channels, or man-made features, such as ditches and extraction pits. Such features may be used unaltered as deposition sites and then be covered with leaves, soil, branches, rubbish, or whatever else is readily available in the local environment. Alternatively, if such hollow features have become infilled previously, then this may provide an area that is easier to dig than the surrounding ground. The offender may not have any awareness of the process of the formation of these hollows and their subsequent infilling but may simply notice an area of softer soil that is easier to dig. However, if the archaeologist can identify potential soft spots in an area of hard geology, then this can provide a targeted focus for ground-level searches that prioritises areas of highest potential. If hollows or soft spots are not available, then the offender might simply deposit the remains at surface level and either make an attempt to cover them over or simply leave them exposed.

In addition to deposition behaviours being influenced by unexpected ground conditions, offenders might also choose deposition sites having never had any intention of digging a grave. A non-comprehensive selection of modes of recent deposition that have not required digging includes:

- Surface-level deposition within woodland: naturally subsumed by soil/leaves/vegetation
- Surface-level deposition: deliberately covered with branches
- Surface-level deposition: covered with soil or other materials
- Surface-level deposition: covered with concrete/built structure
- Disposal in ditches
- Area of tarmac/concrete hardstanding: covered with rubbish
- Railway embankments: surface deposition in vegetation at base of slope
- Dismembered remains: in suitcases within bodies of water, laybys, and wooded areas
- Burnt remains: bodies burnt at ground level
- Bodies dissolved in chemicals and poured away

2.6.1 Conclusion to RAG Mapping Discussion

It should not be assumed that the remains of a missing person have been buried unless specific intelligence suggests that this is the case. This means that the presence of hard geology should not be used as a criterion to assign reduced levels of priority to a search area. Instead, prioritised search areas should be defined using criteria such as intelligence, cell-site and ANPR data, offender profiling, statistical information, reconstructions of the historical landscape, and comparative geology, soil, pollen, and vegetation analyses derived from suspects vehicles, clothing, tools, etc. We can then determine a range of possible deposition scenarios for these areas informed by factors that include the geological character. These will then contribute to the compilation of appropriate strategies for ground-level searches.

2.7 Conclusion

The geospatial sciences as a whole are experiencing a time of considerable developments in technology, data, and analytical approaches. As these factors continue to develop, we are presented with ever-expanding opportunities to develop approaches in forensic geospatial analysis, and it will be important for forensic practitioners to be prepared for this: firstly, in terms of being prepared to identify and harness new technologies and techniques to ensure that our approaches are as well-equipped and up to date as possible and, secondly, to ensure that we are utilising these new technologies and approaches within an experiential framework that enables both the production of accurate and reliable data and the subsequent interpretation of that data according to demonstrable and well-informed rationales.

Chapter 3
Geophysics and Archaeology at Crime Scenes

Pier Matteo Barone

Abstract Irrespective of the type of crime and the modalities of intervention on a crime scene, the approach to the crime scene has recently evolved by virtue of the assistance of forensic geosciences. Studying, knowing, and interpreting the environment and the territory in which the crime took place are essential for providing proper direction to criminal investigations. This is because the forensic geoscientist must be able to search for and collect adequate information from the environment, place it in its specific criminal context, and characterize its actual or presumed narrative.

At first glance, such an approach specifically involves the use of various tools of analysis of various scales, which are completely noninvasive. Geophysics, on a local scale, undoubtedly allows the location and precise mapping of objects (e.g., metal drums or weapons), bodies (e.g., burials), and cavities (e.g., bunkers) of various natures and dimensions, both underground and underwater (generally, in forensic geophysics, the use of electromagnetic instruments, such as GPR or metal detector, is involved). Together with these tools, cadaver dogs are essential to complete the preliminary investigative picture, providing useful information in order to have a further confirmation of the aforementioned geophysical investigation.

The field of forensic archaeology is part of this multivariate approach. The most popular and most requested service involves the use of reading and interpretation techniques of material traces and contexts of the archaeological discipline in the forensic field. In particular, this is the anlysis of the crime, recognition and classification of finds, identification of their origin and time, reconstruction of the spatial arrangement of people or objects in a given place and moment, and the temporal sequence of anthropogenic and natural actions which occurred at the scene of a crime.

Keywords Forensic geophysics · Forensic archaeology · CSI · Non-destructive techniques · GPR · Forensic geoscience

---

P. M. Barone (✉)
American University of Rome, Rome, Italy

Geoscienze Forensi Italia® – Forensic Geoscience Italy, Rome, Italy
e-mail: p.barone@aur.edu

P. M. Barone, W. J. M. Groen (eds.), *Multidisciplinary Approaches to Forensic Archaeology*, Soil Forensics, https://doi.org/10.1007/978-3-319-94397-8_3

Geophysical surveys are based on measurement techniques typical of classical and modern experimental physics, which, in forensic science, can primarily be used for the research of objects and bodies buried in the subsoil or in artifacts. They can also be used for the study of pollutants dispersed in the ground, illegal landfills, hidden rooms, armored shelters, and so on. The correct use of such methods requires, above all, a clarity in terminology that avoids any ambiguity in the definition of the investigative technique. In Italian and international literature dedicated to the application of physical methods to investigative and judicial investigations, there is, in fact, a certain confusion in the use of the terms remote sensing and geophysical exploration. In a broad sense, the term remote sensing refers to measurement methods that acquire information, on a small or large scale, relative to an object or a phenomenon, using appropriate devices that are not in physical contact with the object under study.

With the advent of new disciplines such as forensic anthropology, which deal with the recovery and study of human remains, the multidisciplinary nature of this field is increasingly clear. This multidisciplinary feature emerges mostly when it comes to "finding" and recovering the victim. As already mentioned above, one of the tools for carrying out the search is a cadaver dog. The role of a cadaver dog, until recently nonexistent in Italy, has for decades been a very important component of the human remains' research teams in the Anglo-Saxon countries and Northern Europe.

Archaeology has application in this multidisciplinary forensic field when there is not only a need to search for and locate clandestine burials and recover their remains, but also in the event of concealment of evidence, pollutants, and any other material underground. The sequence of operations leading to the location and recovery of buried remains is divided into three main stages: a preliminary survey of the alleged place of burial in order to determine the location of where exactly the investigation needs to be conducted, the actual excavation phase, and the documentation and recovery of the remains as these are excavated.

The tasks of the archeologist comprise carrying out an excavation, accurately documenting each passage, and ideally recovering the objects found, whether they are human or material remains of another nature. In a forensic context, it may be important to know how to properly remove any human remains. Although, often, in the excavation sites, it is very easy to uncover just scattered bones or teeth that have been picked up and moved by carrion predators. The recovery of remains using forensic archaeology techniques can provide information about the deposition-concealment methods of the corpse and permit the finding of useful clues for the investigative forces.

Archaeological excavation is a destructive operation; consequently, the need to fully document every single advancement has always been a factor of extreme importance. Over the years, more and more accurate methodologies have been developed and perfected with the principal aim of maximizing the amount of information obtainable from the survey site. This approach turns out to be especially reported and useful in the forensic context, where the site of discovery constitutes the scene of a crime and the information stored in it can prove to be of fundamental importance for the reconstruction of the events and the successful outcome of the investigations. Good excavation documentation can be drawn up by suitably trained judicial police officers; however, the analysis of the excavation and the remains, the interpretation of the data, and the coordination of the subsequent recovery operations remain the exclusive prerogative of the specialists in forensic archaeology.

3.1 Forensic Geophysics

Forensic geophysics is the localization and mapping of objects, bodies, or cavities, of various types and sizes, hidden underground or underwater, using specific geophysics tools, for legal purposes. In forensic investigations, a wide range of geophysical techniques are applied that have the potential to verify a contrast in physical properties between a target and the material in which it is buried. Generally, in forensic geophysics, the use of electromagnetic equipment, such as ground-penetrating radar (GPR) and metal detector, is involved, which means it is possible to obtain an approximate image of the variation of physical properties in the first few meters under the surface. Generally, abnormalities in the variations of the physical parameters can be potentially interpreted as coming from buried "foreign" materials. With the help of these techniques, it is therefore possible to identify and precisely define the place of concealment of the target in question, even to the extent of recognizing evidence of human occupation, anthropogenic disturbance of the soil, or soil excavation, both recent and after many years. The geophysical methods have the potential to swiftly and noninvasively investigate extensive areas, where someone has tried to obliterate an underground clandestine burial or, in general, a forensic target.

Forensic geophysics is not only effective for the research and localization of metal weapons or drums, burials, and bunkers, elements that are particularly "visible" and distinguishable from a more or less homogeneous burial or concealment context, but also very useful for identifying areas with high rates of chemical pollution, where the contaminants introduced into the ground alter their physical properties in accordance with their concentration and geometric distribution. A similar discourse can also be made for human burials because the decomposing bodies release slurry, easily identifiable by geophysical techniques as they alter the chemical-physical properties of the surrounding soil. Obviously, as for any discipline in the field of forensic geosciences, surveys of forensic geophysics must be conducted by expert personnel, who are not only aware of the physical principles, but also mindful of the notions suitable for the type of investigation to be performed.

The geophysical approach, therefore, requires specialization and experience. In fact, the development of the instruments, knowledge, and understanding of the subsoils of the area to be investigated and the ability to process and interpret the collected data are fundamental elements for the success of a geophysical campaign. Geophysical methods are very sensitive to small changes in the properties of the earth's material, while maintaining a high resolution for the result.

Field data are acquired through a series of measurements carried out according to regular and parallel profiles, following a very tight mesh previously established, which is linked to the size of the target to be identified.

Once the measurements have been completed, the geophysicist processes the information through specific software. From their analyses, they will be able to produce a map of the detected geophysical anomalies.

The main anomaly that geophysicists must search during their investigations is the so-called "cut" that is present in any material at the time a clandestine excavation is carried out.

These cuts (whether in the ground or in rocky or artificial materials such as rocks or cement) persist for centuries from the moment they are made. This cut is very clear geophysically, due to the difference in chemical-physical properties and compactness between the surrounding material, the cut, and the filling (whether it is carried out with the same type of material or with a different one).

This anomaly, therefore, is the anomaly par excellence. It is always present in cases of clandestine burials, whether they are related to burying bodies, weapons, drums, or any other material. Every good geophysicist should know how to distinguish it in his own prospects.

Naturally, every investigation has its peculiarities, so that sometimes together with the cut, it is possible to recognize anomalies in a geophysical analysis, which are related to the dispersion of slurry or of metal objects such as shell casings and cartridges in the soil.

Understanding the context in which we are operating is therefore extremely important in not only optimizing the choice of methods to be used but also, and above all, focusing on the integrated study of the techniques mentioned above.

While the metal detector is limited solely to the search for metallic objects buried in shallow depths, the GPR—the best geophysical method in the field of forensic investigation—allows detection of anomalies of different types, such as buried bodies, decomposing sludge, pollutants, bunkers, weapons, and any buried object that has a clear variance with the natural geological background. In addition, GPR produces excellent results when used "indoors," i.e., inside homes or huts—built or under construction—to search for secret entrances, bunkers, interspaces, or storage rooms hidden under the floor or behind walls of any material (concrete-armed and not plasterboard, bricks, and so on) (Fig. 3.1).

3.1.1 The GPR: Theory and Methods

GPR, known in the international field by the Anglo-Saxon term of ground-penetrating or probing radar, operatively, involves sending high-frequency electromagnetic pulses (10–3000 MHz) into the ground and measuring the time taken by the emitted signal from the transmitting antenna to return to the receiving one, after being reflected and/or diffracted by possible discontinuities present in the investigated material. The round-trip time (TWT), expressed in nanoseconds (ns), allows

Fig. 3.1 The use of both electromagnetic instruments—GPR and metal detector—in a crime scene

in measuring the distance in time between the antennas and the "target." This distance can be transformed into depth (meters) in the subsoil, if the velocity of propagation of the pulses in the investigated medium can be measured.

The attenuation of these pulses in the subsoil is related to two elements: the presence of moisture in the ground and the chosen frequency. With regard to the presence of humidity, a high level of groundwater runs the risk of not penetrating (or penetrating only partially) the electromagnetic signal, making the ground very conductive.

The choice of what frequency to use depends on the fact that the transmitter is connected to an antenna (Tx), which produces a very short electromagnetic pulse (around 1–10 ns). The duration of the chosen pulse is, in turn, linked to the frequency of the antenna used and to the vertical resolution required, i.e., the ability to distinguish between two layers or objects close to each other. In other words, the higher the antenna frequency, the shorter the pulse, which translates into a low signal penetration (because the attenuation also depends on the frequency) but in a higher vertical resolution.

The GPR instrumentation has two possible configurations: the so-called bistatic configuration, in which the transmitting antenna is physically separated from the receiving one, and the monostatic configuration, in which the transmitting and receiving antenna coincide.

The graphical representation of GPR data is a fundamental step for understanding and interpreting results. These results report grayscale radiations (or stratigraphy), and modern software allows a very high visual resolution and definition. Moreover, if the acquisitions have provided parallel profiles inside a grid, it is possible to obtain and therefore display maps (i.e., floor plans) of the investigated area, which not only represent the geometries of buried objects at various depths, but also the dimensions, normally using an average envelope algorithm, also known as the average envelope amplitude (Fig. 3.2).

To correctly interpret a radargram, it is necessary to know how the section was acquired. The pulse transmitted by the radar antenna does not propagate in the ground or in the material in a precise manner like a laser but behaves like a so-called radiation cone, "illuminating" the buried target even before being perpendicularly above the target itself (like a lamp lit in the darkness of a room). The diameter of this cone increases with the increasing depth of the GPR signal. Moreover, its dimensions also depend on the conditions of the acquisition surface and the frequency of the antennas used (e.g., high frequencies restrict the diameter of the cone) (Fig. 3.3).

The presence of a void in the subsoil or of any specific object more or less produces a characteristic electromagnetic response: the diffraction hyperbola. The hyperbolic anomaly results from the reflection of the source point (buried target) and occurs, as we have seen, because the energy is emitted in the form of a cone, which "illuminates" a larger portion of the target itself. Consequently, the signal

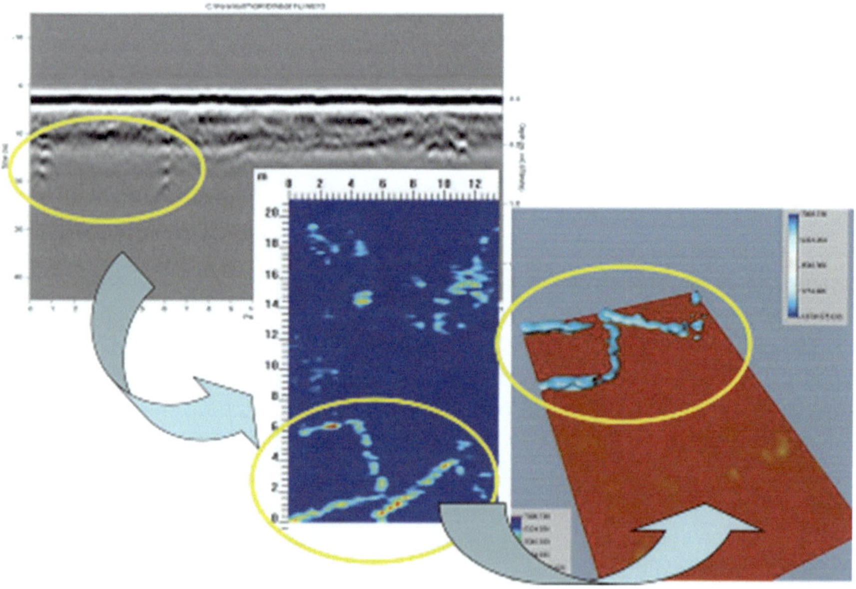

Fig. 3.2 Exemplification of how the same anomaly (in this case, a wall structure) can be visualized in three different ways: a radargram or stratigraphy of the subsoil (top, left), a map (or planimetry) at different depths in the subsoil (in the center), and a 3D reconstruction (below, on the right) acquired with GPR

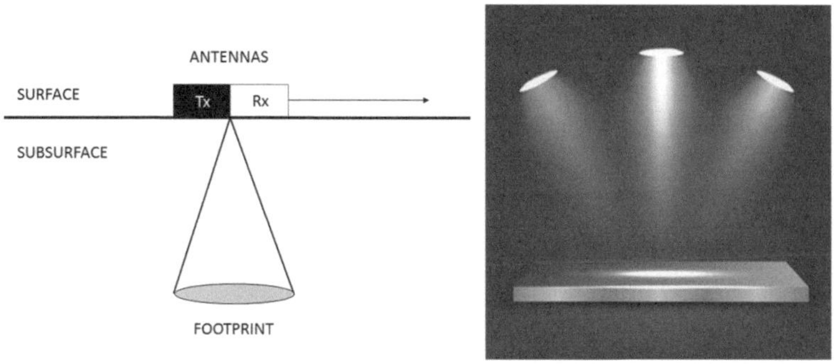

Fig. 3.3 The transmitting antenna (Tx) emits a signal that does not travel vertically in depth like a laser but creates a cone of radiation that "illuminates" the target and is reflected in the receiving antenna (Rx). On the right is the analogy with a lighting cone, for the purpose of providing an example

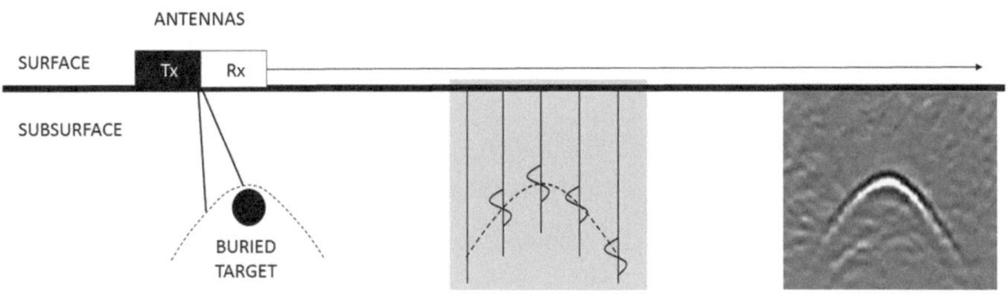

Fig. 3.4 The diffraction hyperbola, visible in this figure, derives from the reflection of the intercepted target. It should be noted that only the apex of the hyperbola identifies the correct position of the buried object

is reflected not only from the target directly perpendicular to below the antennas, but also shortly before and shortly thereafter, due to the transmission of oblique waves. Only the apex of the hyperbola corresponds to the actual position of the source (Fig. 3.4).

The maximum horizontal resolution is roughly the impression of the radiation cone (or illumination area). TWT of the signal, and consequently the estimate of the depth, can be calculated by what is known as the calibration of hyperbolic queues derived from an anomaly. Important to emphasize, however, is that it is possible to determine the depth of a target as long as the speed of the signal penetration in the material or materials is known.

With the exception of buried conductive materials (e.g., metal, which has a high conductivity and magnetic permittivity), electromagnetic waves pass through the buried target, continuing their penetration and producing different reflections at different depths. In some cases, this effect allows the estimation of not only the depth of the upper part (top) of the object but also its vertical dimensions (e.g., in the presence of an underground tunnel, it is possible to identify not only the top of the tunnel but also the bottom of the same).

3.2 Cadaver Dogs

Cadaver dogs (or the international term K9)[1] have the primary task of locating the remains of missing persons. Despite extensive research in the field of technology, it has not been possible to find a worthy substitute for the dog, who has been used in these operations since 1974. In fact, on this day, the first dog, which was used for searching corpses, was trained by the New York Police Department. After ascertaining that dogs could prove to be a useful tool, the FBI, too, began breeding dogs for this type of work, including some volunteers from the National Search and Rescue Association.

The use of cadaver dogs is based on the extraordinary sense of smell of such animals. In a medium-sized dog, the total surface of the olfactory mucosa is 150 cm^2 compared to just 5 cm^2 in an adult man, and it is a tenth of a millimeter thick against 5 thousandths of a millimeter of man. There are about 225 million canine olfactory cells compared to 15 million in a human being; consequently, the dog has a 500/700 times greater capacity to perceive odors. The dog can perceive the smell of 2 mg of blood serum or even 5 mg of urine and can sense acetic acid in aqueous solution in a ratio of 1 part in 1 million.

In practice, these animals, by virtue of their particularly developed nose and a specific training program, can recognize the chemical substances (about a 100) characteristically released by a cadaver.

To ensure that the smell can be detected and therefore communicated, it is necessary that it exceeds a certain threshold and also remains in the dog's presence for a sufficient period of time. In other words, it is necessary that the odor exceeds the stimulus threshold, has an elevated period of exposure or is sufficiently intense to trigger recognition, and remains for a sufficient period of time for the dog to pick up and communicate. The training of dogs is based on these concepts, which involve the use of "baits" of odorous material. These materials may consist of pork, blood-soaked pieces, human teeth, or chemical substitutes such as cadaverine and putrescine.

The search for a corpse is configured as a hunt for the dog, where the dog behaves as a follower and the operator is the leader; locating the corpse is therefore understood as the capture of the prey, whose division is reproduced using a game preferred by the operator. It is clear that this activity requires a well-established relationship between man and animal, which is developed over the course of many years. From the time they are puppies, the operator can begin an initial research training by hiding the toys of the dog or accustoming the dog to look for things or objects. In Italy, the basic training is based on the activity involving buried pork first and then human organic material, wherein small holes are made in the ground about 3 cm in diameter, at a depth of about 10 m, at a distance of about 30 cm from each other, in a straight line for 10 m, and on more rows 15 cm of each other.

---

[1] Symbol linked to the assonance between the Anglo-Saxon pronunciation of the letter "K" and the number "nine" and the English pronunciation of the word "canine".

This type of signaling also requires a special type of training. Most Italian dogs are trained to report by stopping and sitting in front of the source of smell and waiting for gratification; this behavior constitutes the wait for the "jump of the rabbit" or for the localized prey to exit the lair. The operator, without being seen by the dog, places it at the point established by the dog, creating the effect of gratification. Alternatively, the dog can scratch at the ground, which is behavior aimed at checking if the smell increases with the removal of the soil. Among more communicative dogs, a behavior is established, which is aimed at alerting the operator by traveling between the signaled location and the position of the operator. In some cases, it is possible to conduct surveys via T-bar[2] in the ground in order to allow a greater outflow of the smell and increased likelihood of identification.

From this, it is clear that the dog alone cannot resolve all problems, as it requires great collaboration and communication between the investigation departments and a specialized preparation by the operator.

When conducting research, for example, it must be taken into account that the concentration time of the dog in a targeted exercise is reduced to 10/15 min. Therefore, the research area should be limited to the real possibilities that the dog meets the source of the smell, so that it can be minutely inspected by the animal. On open ground, the area is divided into corridors of about 500 m^2, arranged perpendicular to the wind direction. Each of these corridors will be considered as sectors and classified with numbers or letters.

In some very rare and particular cases, the cadaver dog is so well-trained and has reached such a level of sophistication that it senses the difference between human and animal organic matter. With the help of the conditioning method used during the training of dogs, they are able to recognize any chemical substance that is fixed in their olfactory memory by their trainer (this is the case, e.g., of dogs trained by the Cynthia Unit of the State Police of Milan Malpensa).

Finally, it should be noted that these dogs are also able to detect the presence of a body submerged in water. The dogs do not dive as contact with the water surface is enough to indicate nearness to the source. From the boat, they are already able to pick up on these volatile particles released by the decomposing human body. Maybe they are on standby all the time, or they doze off, but when they enter the field of the odor and perceive the possibility of identifying the chemicals they are trained to recognize, they immediately give the signal that something is there at that point.

On most occasions, a dog is the most effective and efficient resource. In combination with other technical resources, a dog can provide an exact indication of where to begin the excavations, thus safeguarding the integrity of the finds to be recovered.

[2] The T-bars are T-shaped rods with a length of about 100 cm, with a diameter of 2.5 cm, inside cavities with an opening of 40 cm on the side, also used in archaeology to verify the consistency of the ground in depth for the search for anomalies.

3.3 Forensic Archaeology

Archaeology is the study of the human past, ancient, and recent, through material remains. From the fossil remains of millions of years ago, which belonged to our first human ancestors in Africa, up to today's twentieth century buildings, archaeology analyzes the physical remains of the past in search of a broad and global vision of human culture.

Forensic archaeology is the application of archaeological principles and methodologies to identify and recover evidence within the confines of the criminal justice system or for use in civil cases, operating as a discipline in its own right.

People authorized to conduct forensic archaeological investigations are mostly trained in traditional archaeological techniques but are flexible enough in their approach to adapt these methods to the different forensic contexts that may arise in criminal or civil proceedings. Integrated skills and considerations outside the parameters of traditional archaeology are the possession of a basic knowledge of law enforcement and legal procedures (as well as the ability to cooperate productively with law enforcement personnel), ability to efficiently and effectively conduct investigations under tight deadlines and under the attention of the media, as well as the ability to navigate skillfully in situations that deviate from traditional archaeological experience, such as in the presence of a burial that includes conservation of the remains of soft tissues.

The sequence of operations leading to the location and recovery of buried remains is divided into the following three main stages: a preliminary survey of the alleged place of burial in order to determine the exact location where the investigation is to be performed, the actual excavation phase, and the documentation and recovery of the remains as they are excavated.

The survey, contextualization, and localization provide for the interpretation of the territory in order to search for a presumable burial site. They are carried out with the help of various instruments, both large- and small scale. For example, aerial photographs can easily provide important information about the area to be investigated (abnormal changes in vegetation cover over large areas or changes in soil morphology), and their use in the preliminary stages of a survey can be effective, leading to a significant saving of time and energy. As seen in the previous paragraphs, on a local scale, geophysical techniques and K9 units are applied, which allow a more precise localization of the buried target.

The tasks of the archaeologist consist of carrying out an excavation, accurately documenting each passage, and ideally recovering the objects found, whether they are human or material remains of another nature. In a forensic context, it may be important to know how to properly remove any human remains. Although, often, at excavation sites it is very easy to just uncover scattered bones or teeth that have been picked up and moved by carrion predators.

The recovery of remains using forensic archaeology techniques can provide information about the deposition-concealment methods of the corpse and permit the gathering of useful clues for the investigative forces. Analysis of the context, excavation and remains, interpretation of the environmental data, and the coordination of the subsequent recovery operations remain the exclusive prerogative of specialists in archaeology. Archaeology is a destructive process. For this reason, it is preferred to integrate it with other methods in order to pass from a nondestructive analysis to a more invasive one. This minimizes the loss of information.

For this reason, three basic objectives can be summarized in an archaeological forensic investigation. The first is the understanding and interpretation of taphonomic events—the history of a site after it has been created through the deposition of remains. Taphonomic events are both natural (N-transformations) and cultural (C-transformations) processes that occur at a site and alter or transform it over time. These processes include N-transformations, such as seasonal water runoff, high levels of animal or insect activity, and tree root activity and growth, and C-transformations, such as informal excavation (due to the lack of awareness that the position is a location of criminal activity or action of a civilian), deposit of waste or other objects unrelated to the deposition event, and high levels of human traffic that can disturb the original deposition contexts. Understanding and being able to interpret the transformations that these processes cause at a site are essential for an accurate archaeological investigation (Fig. 3.5).

The second main objective is the reconstruction of the events that have caused and occurred with the creation of the site and the body deposition. This is accomplished through the collection of surface disturbance elements, excavation methods, and detailed and photographic documentation at all stages of the survey. It must be remembered that the archaeological forensic investigation of a site is inherently destructive and forensic archaeologists must therefore take all precautions to preserve the maximum amount of evidence and potential contextual information (Fig. 3.6).

Fig. 3.5 Some phases of archaeological excavation during a forensic investigation

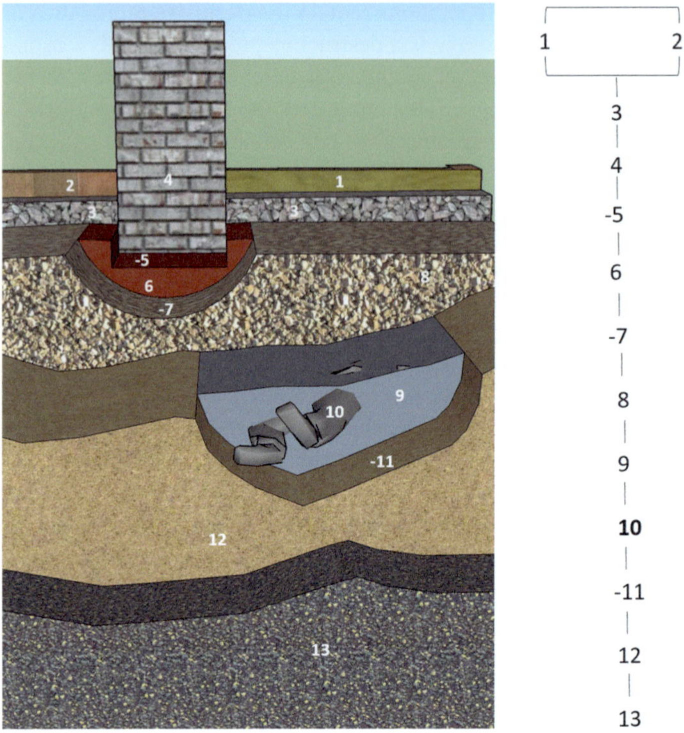

Fig. 3.6 Section expanded in 3D format of a forensic archaeological stratigraphy and related Harris matrix. Note how the cuts made in the ground are indicated with the sign "-," while the US 10 is in bold because it indicates the buried body

Finally, based on the evidence that the recording of this material provides for the reconstruction and the taphonomic history of the site, an investigator is able to achieve the third objective, which is a conscious interpretation of the events surrounding the deposition of the deceased person, which will help in resolving the case.

It is therefore evident that archaeologists specialized in forensic investigations provide the police with the complete reconstruction of a crime scene, with the possibility of investigating cases that are often very distant from each other and crime scenes that are still "warm" and very recent but maintaining the same high standards of evidence and resolution recovery.

3.4 Case Reports

3.4.1 "A More Efficient Search"

"It's about creating a more efficient search," so Jon Dittmar (ERA Technology) commented on one of the recent investigations that witnessed a close collaboration between geophysics (specifically GPR) and forensic geo-archaeology.

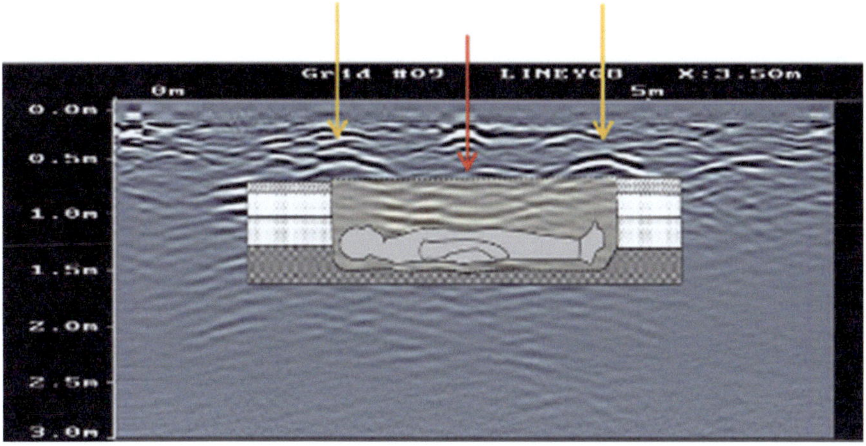

Fig. 3.7 In the radargram, yellow arrows indicate hyperboles linked to the cut made in the ground to obtain the burial, while the red arrow indicates the horizontal reflector—the filling after the cut and what remains of the body, with chemical-physical characteristics different from the surrounding soil

In November 2007, in the garden of a house on Irvine Drive in Margate, Kent, England, a long investigation seemed to have come to an end. Fundamental to the research was a GPR system that uncovered the bodies of two teenagers, Vicky Hamilton and Dinah McNicol, who disappeared in 1991. The geophysical system, in fact, helped the forensic investigators and the police in identifying the bodies of the two girls.

Jon Dittmar is an engineer at ERA Technology, the company that, for the first time in 1994, helped forensic investigators and police in finding bodies in Fred and Rose West's home on Cromwell Road, Gloucester. Dittmar is not only aware of the potential and ductility of GPR, but also, specifically, recognizes the absolute effectiveness of the close collaboration between geophysics and other forensic geosciences.

But how can a geophysical expert understand, analyze, and interpret an anomaly as a possible underground burial? In this case, GPR in real time obtains an electromagnetic response that is configured, as shown in Fig. 3.7. Clearly, every case and every terrain have their characteristics, but generally the presence of a hyperbolic event, a horizontal reflector, and another hyperbolic event is an index of a break, a "cut" made in the ground and subsequently filled, leaving an indelible and lasting mark in time for centuries to come (as taught in archaeology).

3.4.2 Weapons and Caches

The scientific investigations on the crime scene have brought to the fore the geophysical skills (particularly GPR) that are required for determining the locations and identifying missing persons, who are buried in the ground and beyond. Recently,

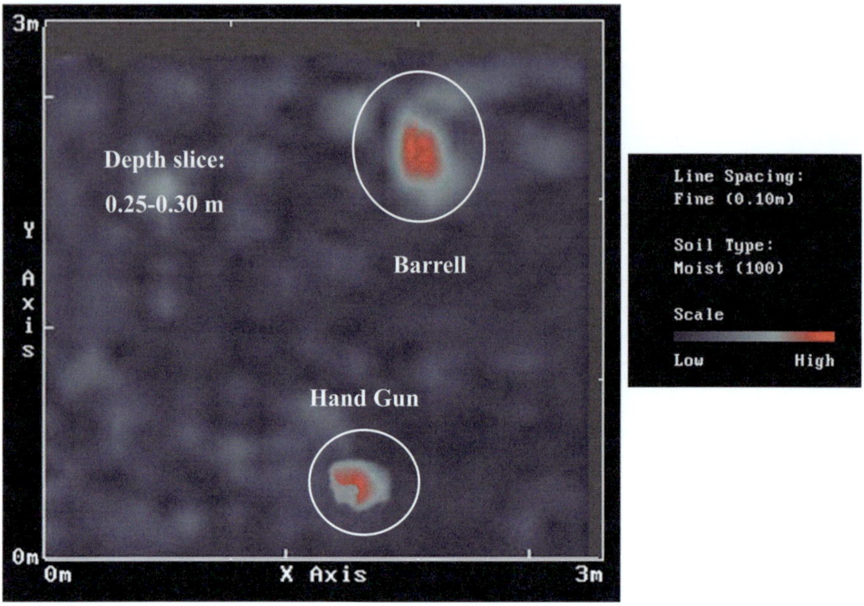

Fig. 3.8 GPR map to the depth of about 30 cm in which the presence of a gun and a barrel is clearly indicated

in fact, this method has been tested to understand how accurately it can determine an object, even of small dimensions, buried in the subsoil.

Figure 3.8 displays the results following a scientific investigation in search of a gun illegally disposed of at a shallow depth in the ground and a plastic barrel with drugs and money. It illustrates how the GPR map not only indicates the clear location of the object, but also its geometry, although the gun was buried by itself, of a small size, and rolled up in a plastic bag. As for the barrel, given its size and the partial presence of air inside (an element that stands out very much in an electromagnetic response), its position and dimensions are very clear.

3.4.3 GPR, K9, and Archaeology

There are many cases in Italy and abroad in which cadaver dogs accompanied forensic investigations using GPR. Well-known among them is the British case of Madeleine McCann and the Italian case of Marina Arduini. Yet such concerted actions could be greater if the experts involved in these investigations put in excellent and undisputed groundwork.

A couple of years ago, the Geoscienze Forensi Italia team[3] was involved in the reopening of a cold case in Marcianise (CE), which involved the disappearance of

[3] Geoscienze Forensi Italia is an Italian platform of various experts in the various fields of forensic geosciences (geology, pedology, mineralogy, archaeology, geophysics, geochemistry, remote

Fig. 3.9 A photograph of the joint investigation of GPR/K9 operations

Pasqualino Porfidia. This case caused quite a stir at the time (1990) and was aired on television programs such as “Chi l’ha visto” (“Has anyone seen them”). The reopening of the case, based on new indications, required the joint use of GPR and cadaver dogs in the preliminary phase to look for nondestructive and precise anomalies related to burials or corpse occultations, which then allowed a more detailed research through an eventual archaeological excavation. This integrated survey assisted in locating a specific area in which both GPR and the cadaver dog, independently, provided a positive response (Fig. 3.9). The subsequent archaeological excavation then only partially confirmed this positive identification.

sensing, geotechnics) that provides services to scientific investigations in the forensic field with commitment, professionalism, and experience (www.geoscienzeforensiitalia.com).

3.5 Criticality of the Investigation

The geophysical approach of archaeology requires specialization and experience. Indeed, the development of the instruments, knowledge and understanding of the subsoils in the area, as well as the ability to process and interpret the collected data are fundamental elements for the success of a geophysical endeavor. If the acquisition of the given GPR may seem trivial (in reality, it is both complex and delicate), its interpretation requires years of experience and innumerable studies.

Furthermore, those who commission this geophysical survey (whether it is a judge, police official, public prosecutor, or lawyer) must not only be aware about the potential of the instrument but also its limits. This is not a magic wand, but a scientific instrument that has uncertainties in measurement that must be considered when the data is acquired and interpreted.

Likewise, cadaver dogs, although a valuable resource, are living beings and as such are sometimes inclined to err when used in an erroneous manner. Moreover, their full potential is yet to be explored. Nevertheless, it should be recognized that the success of the search operations also depends on the organization of activities, which must take into account fatigue and potential distractions for the dog.

Finally, we must consider that every human operation is selective, both voluntary and involuntary, and for this reason, the moments of excavation and, more importantly, documentation are the phases involving maximum loss of information, no matter how extensive the data recovery equipment might be (Fig. 3.10). In this respect, we need to have a very precise strategy and pay meticulous attention to the

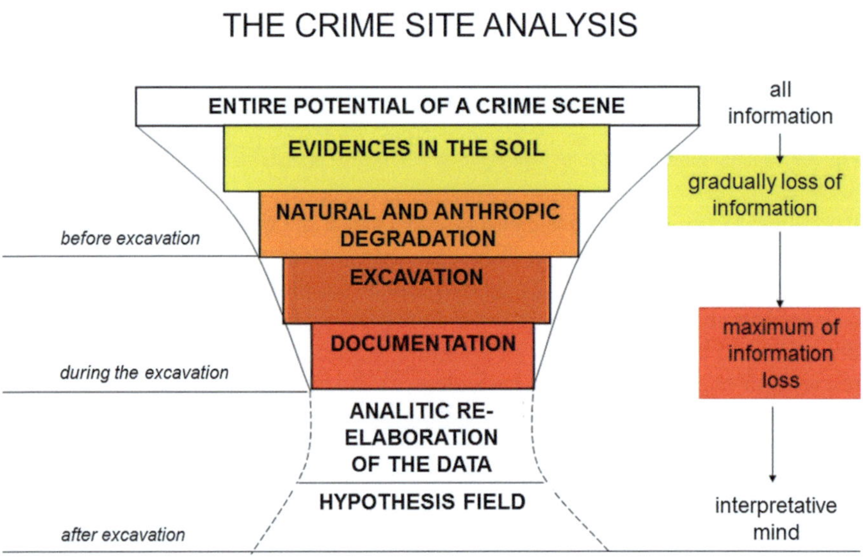

Fig. 3.10 Graph of the forensic archaeological crime site analysis

field to reach the maximum expansion of the data in our possession. This happens through re-elaboration and the formulation of reconstructive hypotheses in order to transform the sediment into history.

The origin and the context are extremely important, without which no evidence can be compiled for arriving at a clear and complete representation of the act of deposition, the events surrounding it, and the identification of people involved. Once all the evidence from a site has been recovered and treated according to the correct chain of command, a site is typically covered (filled with the soil that was originally excavated by the investigators). However, despite the filling, a slight scar is often visible at the excavation site in the form of a depression in the earth. This is because not all of the soil and material removed is replaced nor is it replaced with the same orientation or compaction that originally characterized it.

The coordination between different dogs and activity planning among other figures present onsite, such as the geophysicist, geologist, forensic archaeologist, or the medical examiner, are issues of fundamental importance.

Chapter 4
Forensic Geophysics and the Search of Building Interiors, Peat Bogs and Freshwater

Alastair Ruffell and Laurance Donnelly

Abstract Geophysics is one of the assets commonly deployed in the multi-proxy search for targets buried in the ground and concealed in water, most especially associated with criminal activity (e.g. human remains, graves, weapons/explosives/contraband, toxic waste). Here, we review and provide new case studies in three environments: (1) the search for objects inside human-made structures, (2) the search for buried homicide victims and human remains at unknown locations in peat bogs, and (3) the use of water-penetrating radar (WPR) in the detection of human remains in water. The latter section is expanded to the use of WPR as a reconnaissance tool in mapping areas of thickened sediment fill in water bodies, as a possible search area for sunken and then sediment-buried objects. We introduce a new term – 'sinkability' – to convey the concept of subaqueous areas of soft sediment where objects such as human cadavers could reside below the sediment surface.

Keywords Geophysics · Forensic search · Freshwater · Peat bogs · Homicide

4.1 Introduction

The use of geophysics in the search for buried or water-submerged items is now well-established (Buck 2003; France et al. 1997; Nobes 1999, 2000). Early work on the detection of buried cadavers concentrated on the use of ground-penetrating radar (GPR; e.g. Strongman 1992). Subsequent workers advocated a multi-proxy approach for the use of geophysics (Ellwood et al. 1994) or highlighted the use of other methods such as resistivity (Pringle and Jervis 2010). Due to the variable success/failure of all search methods, two themes have emerged: the first being the deployment of multiple search assets (including geophysics; Harrison and Donnelly

A. Ruffell (✉)
School of the Natural Built Environment, Queen's University, Belfast, N. Ireland, UK
e-mail: a.ruffell@qub.ac.uk

L. Donnelly
International Union of Geological Sciences, Initiative on Forensic Geology, Lancashire, UK

P. M. Barone, W. J. M. Groen (eds.), *Multidisciplinary Approaches to Forensic Archaeology*, Soil Forensics, https://doi.org/10.1007/978-3-319-94397-8_4

2008) and the second being the use of a desktop study (including an evaluation of the geology, soils, past land use, depth/nature of target, etc.) to choose the most appropriate or 'fit for purpose' geophysical technique to use in the search for buried objects (Donnelly and Harrison 2010; Donnelly 2013a, b; Pringle et al. 2012). Such a desktop study is one of the fundamental principles for the correct application of geophysics (Reynolds 2011). Forensic geophysics had early advocates of GPR (Strongman 1992), followed by electromagnetics or EM (Fenning and Donnelly 2004), resistivity (Pringle et al. 2016) and others: each technique has its own benefits yet rarely is the single solution to finding the target. An exception may be the use of metal detectors (in areas with no other metal) for the location of buried guns, knives, coins, metal boxes, etc. The development and applications of the conceptual geological model (CGM) alongside the evaluation of diggability and the applications of the Red-Amber-Green (RAG) prioritisation system were significant advances in the UK, Italy and elsewhere, for police and law enforcement ground searches for burials (Donnelly 2003, 2013a, b). This, for the first time, enabled police searchers to consider the ground conditions (geology) and the characteristics and properties of the target being searched for so that the most suitable and practicable search asset could then be determined, which frequently included (and now includes) geophysics. In terms of geophysics, this approach facilitated the choice of geophysical instruments that were most likely to provide a high assurance for the presence or absence of a suspected buried target (Donnelly and Harrison 2017). Geological trace evidence underwent a similar evolution, from 2004, when many published articles appeared advocating a specific method (see the chapters in Pye and Croft 2004 for examples), when in reality, a multi-proxy approach to the analysis of something like soil is recommended (Rawlins et al. 2006). Through this evolution, geophysics now has its place in the range of assets available to the forensic geologist when trying to locate a target for excavation and by the forensic archaeologist in the forensic recovery of a buried target. A review of the geophysical methods is available in Pringle et al. (2012), and there is no need to repeat the range and applicability of techniques described in that review. The same comment may be applied to the forensic-based search of water bodies, where individual methods maybe appropriate when conditions determine this but where the multiple use of a desktop study, a range of geophysics, as well as hydrology, detector dogs and divers, is considered the best practice (Ruffell et al. 2017). In reality, searches for items associated with criminal activity or humanitarian rescue are usually time-limited, so the more rapid methods of GPR (Barone 2016; Barone et al. 2016a, b) and EM/metal detection (Pringle et al. 2016; Donnelly and Harrison 2013) have gained significant popularity.

4.2 Recent Advances and Debates

The place forensic geophysics has in a search is now well-established, being based on a desktop study to choose the most 'fit for purpose' technique (Donnelly and Harrison 2013) and as part of a range of methods available to those conducting the search (see above). However, misconceptions still remain, and these are considered here (1) the use of geophysics on, inside and under buildings/structures, (2) the use of GPR/WPR in peatlands/bogs and (3) the geophysical detection of human tissue in freshwater using WPR.

4.3 The Use of Geophysics on, Inside and Under Buildings/Structures

Although this was considered by Ruffell et al. (2014) from a search perspective, comments from co-workers at recent conferences indicate that there is a misconception amongst non-geophysicists that most techniques comprise some kind of frame- or carted-mounted apparatus that is moved around a field or a garden. That sideway-looking geophysics is possible, and thus imaging into walls is perhaps not so apparent to non-practitioners, and if it is, then GPR or metal detectors (EM) tend to be what is thought of. This is fine, except most of the ground-based geophysical methods available can also be used on vertical surfaces, for instance, resistivity has been effectively used by Tsourlos and Tsokas (2011) and Mol and Viles (2013) to image water ingress and (most especially for forensic work) voids inside structures and behind walls. Similarly, ultrasonic surveying was combined with GPR by Cassidy et al. (2011) in order to try and overcome the problems of metal rebars in concrete structures. Another way of using GPR to overcome the issue of metal in structures, or to try and image non-metallic features in concrete, is to survey from the opposite side to the interference, just as it is informative to the non-specialist to hear about sideway-looking geophysical imaging (Fig. 4.1a), so upside-down methods can be used, described below. Common forensic searches may be focussed on objects buried under roads but most especially in the concrete of bridges or beneath newly constructed building foundations, presumably to hamper the authorities wishing to excavate. While it is not easy to close a thoroughfare such as a road or pavement for a survey and excavation, this is more practicable than doing the same on (and in) a concrete bridge. Ruffell and McKinley (2008a, b, p. 290) recount how the location of a well-known missing person in North America (the notorious Teamster Jimmy Hoffa) was suspected to be under a tarmac in a garage forecourt in Massachusetts. Hoffa has variously been suggested as buried inside the concrete of the Brooklyn Bridge or in a farmland near Ohio or Illinois and fed to alligators in Florida. Likewise, some of the victims of the East End gangsters, the Kray brothers, are thought to be in the concrete of the Stratford Flyover, London (*op cit.*). Figure 4.1b shows how, if rebars are present in the bridge deck, a technique such as ultrasonic (Cassidy et al. 2011) or in this case high-frequency GPR may be used upside-down to image the inside of the bridge deck from the reverse side. In the case study presented (Fig. 4.1), non-metallic (basalt fibre; see Yeboah et al. 2013) rebars were imaged using different high-frequency GPR antennas. The 3D outputs are similar to those derived from both conventional ground-penetrating radar on structures (Cassidy et al. 2011) and those from sideway-looking radar (Fig. 4.1a, after

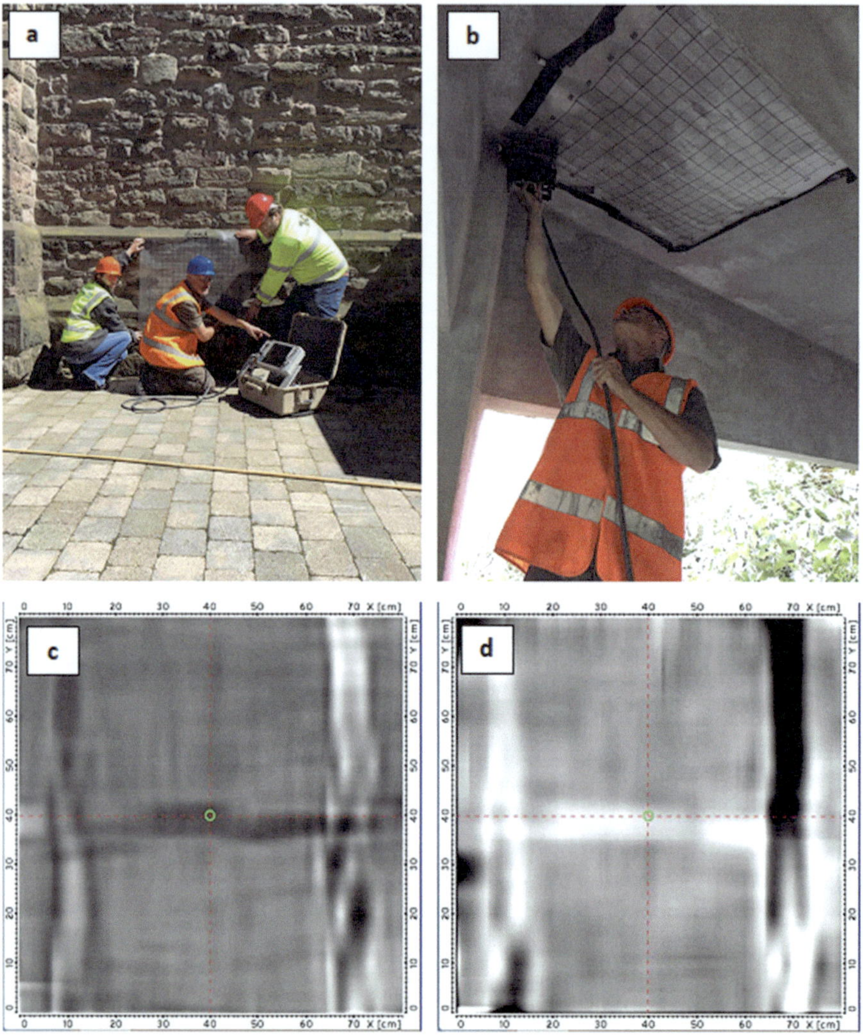

Fig. 4.1 Sideways and upside-down-looking GPR. (**a**) Example of the use of sideways-looking radar to detect voids in an historic masonry wall (Adapted from Johnston et al. 2018) using 1.2 GHz, 1.6 GHz and 2.3 GHz antennas, courtesy of Mike Langton (Mala Geoscience). (**b**) Use of upside-down GPR to image a modern concrete bridge, with basalt-fibre rebars using 1.2 GHz and (this image) and 2.3 GHz antennas, results in (**a**, **c** and **d**). (**c**) Mapped 2.3 GHz output from the survey shown in (**b**), at 10 cm depth, with basalt-fibre rebars. (**d**) Mapped 1.2 GHz output from the survey shown in (**b**), at 15 cm depth, with basalt-fibre rebars (Yeboah et al. 2013). Note the polarity phase change (white/black) with different depths/antennas in (**c**, **d**)

Johnston et al. 2018), which belies the fact that gaining such data is physically demanding, which is probably why this part of the chapter is the first publication on upside-down GPR for forensic searches.

4.4 The Use of Geophysics in Peat (Bogs)

Areas covered in peat (also known as 'peatlands' or 'bogs') comprise organic-rich to purely organic soils that form in temperate humid climates (Hobbs 1986). They are frequently nonsaline water-logged environments that may contain 90% water and occur throughout Northwest Europe (such as upland Scandinavia, the British Isles and countries with low relief such as the Netherlands), North America (especially Canada and Alaska), upland regions of South America (Colombia, Peru, southern Chile/Argentina) and New Zealand. Peatlands are commonly remote, with large expanses of homogenous landscape. Peat is easily diggable which facilitates an offender's ability to dig a grave and bury a person or item in a relatively short time (sensu Harrison and Donnelly 2008; Donnelly and Harrison 2013). Their water-logged nature means that some of the techniques used in the search of freshwater are applied, should the target be suitable (magnetometers, GPR, EM, detector dogs). An advantage when searching on peat is that some areas may be surveyed by walking on them: the disadvantages are that peatlands/moors are dynamic environments, subject to rapid change (bogslides) and excavation is extremely difficult/hazardous due the collapse of the sidewalls and water ingress. The location and diggability of peat have resulted in peatlands being the focus of searches for items of forensic interest in places such as Northwest Europe (northern England, Wales, Scotland and Ireland). In Ireland, the work of the Independent Commission for the Location of Victims' Remains is most pertinent (ICLVR; see Knupfer et al. 2018; Chap. 15 this volume), while Colombia (Molina et al. 2016) describes the use of different geophysical methods in imaging the subsurface for experimentally buried pig cadavers.

Searches may be for weapons, drugs, contraband and, of interest here, human remains in unmarked, shallow, graves. The latter, if buried without metallic objects or clothing, are exceptionally challenging to locate, and the efficacy of GPR in imaging either human/animal burials or non-metallic items (clothing, sacks, bags, ropes, plastic containers) in peat has been the subject of debate at forensic geology and archaeology conferences and other meetings worldwide. In this study, items that replicate those worn by a missing person in northern England were buried in a proximal location to the suspected grave. The case concerns the abduction by serial killers who buried their victims in the same area and in highly comparable geological conditions to our test site. The significance of establishing this control location was to (a) provide the opportunity for the police search teams to become familiar with the operation and deployment of the GPR, magnetometer and electromagnetic instruments in the same geological setting but outside the search area cordon, (b) verify that the GPR and other geophysical instruments were operational and the detection (depth) limitations could be identified (e.g. the target could be buried beyond the depth capable of being detected with the device being tested), and (c) check that the GPR and other geophysical instruments were functional at the start and end of each search phase to provide a high assurance that the target was not present if no anomalies were detected. This case study was undertaken in an area where peat dominates. The peat was observed to contain pipes and voids that facilitate the flow of leachate from the target (Donnelly 2003, 2008). The basal contact of the peat comprises boulders at the interface with the underlying periglacial deposits and derived from in situ weathering of Namurian (Carboniferous) sandstone. This was problematic as the GPR anomalies often were associated with voids and boulders and therefore not the buried target. A detailed understanding of the geology and geomorphological processes from the desktop study, fieldwork and experience is therefore important before GPR is deployed. The blind use of geophysics in such a location could identify natural features such as the peat pipes and boulders as potential forensic targets.

Details of the items worn by the missing person were reported by family members to the police: near-identical items were buried at the control site established a separate target comprised a similar spade. Initial GPR surveys undertaken in the mid-1990s and repeated in the early 2000s were unable to detect the buried control clothing. However, these targets were detectable with ease when tests were repeated in 2015 and 2016. This is not fully understood; however, it is possible that changes in the engineering geological properties and geotechnical characteristics of the reinstated peat after digging and burial of test items and advances in GPR instrumentation enabled the targets to be located. Two generations of GPR antenna were deployed over the control site, both manufactured by Mala Geoscience (Sweden) (Figs. 4.2 and 4.3).

The results show how both the older (manufactured 2006) and newer (manufactured 2016) antennas imaged the clothing and (less surprisingly given the metal content) the spade. A criticism of this experiment is naturally that the location of the targets was known, begging the question as to whether a speculative search of a larger area would have shown anomalies consistent with the homicide victim or indeed the actual items simulated here. The answer lies in the fact that GPR would not be used as the only search tool in such a location as a large upland peat bog but rather be both focussed using the case intelligence, desktop study information and in situ evaluation of the geology/diggability (see above, e.g. Donnelly and Harrison 2013) and used in conjunction with other geophysical methods (e.g. electromagnetics) and other search assets (probing, detector dogs, GIS-based viewshed analysis). Nonetheless, multiple anomalies would be predicted from other buried objects, when these would be prioritised (based on location, geophysical signature, etc.) for the probing and deployment of detector dogs, maybe followed by exploratory invasive excavation.

A second observation by the authors has concerned the poor results from GPR on recent (0–40 years old) homicide burials in peat, yet the overwhelming success when using GPR in the search for unmarked burials resulting from the Irish Potato Famine (1835–1852), both in peat bogs and elsewhere (Ruffell and McAllister

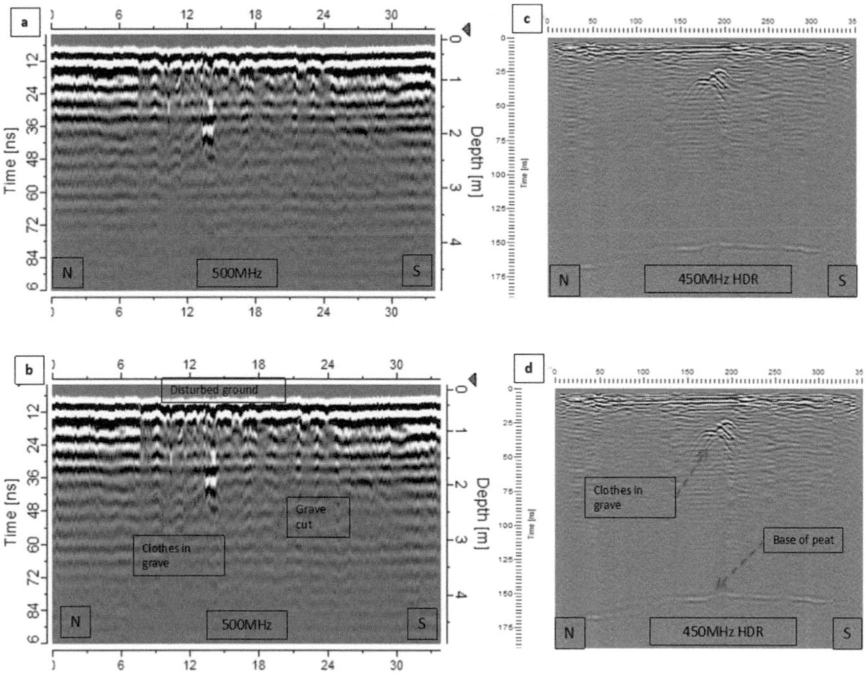

Fig. 4.2 GPR in an upland peat bog, northern England (UK), showing a comparison of an older (manufactured 2006) 500 MHz shielded GPR antenna (Mala Geoscience) and a newer (manufactured 2016) 450 MHz, high dynamic range (HDR) antenna (the latter courtesy of Mike Langton, Mala Geoscience). All data are unprocessed. (**a**) 500 MHz data across a test location where clothing that replicated an actual homicide was buried some 10 years previously in a test pit, dug to emulate a homicide grave. Data gathered in October 2015. (**b**) Interpretation of the data shown in (**a**). (**c**) 450 MHz HDR GPR data across the same clothing burial location shown in (**a**, **b**). Data gathered in November 2016. (**d**) Interpretation of the data shown in (**c**). 500 MHz and 450 MHz data were also collected across a simulated dug grave of the same age, with no contained objects and no GPR anomalies (data available on request)

2014). Jonny Geber (pers comm., 2016) has shown through his work (see Hilts 2013) that victims of the famine were commonly placed in a shroud, with or without a pine coffin, and placed in individual or mass burials, with lime (CaO, calcium oxide) added to both stifle odours and disinfect the site and supress the spread of cholera. A well-known example of this (for the general reader) is the closing scene of the film Amadeus, where lime is thrown on the unmarked grave of Mozart for the same reasons. We conjecture that in the wet climate of Ireland, and especially in the subsurface of peat, this CaO reacted with water to harden and thus create a carbonate carapace over the burials. The success of GPR in surveying such ground may be down to the supra-carbonate cover and underlying burial airgap, providing a good geophysical target (Fig. 4.4). Work by the authors at the homicide burial test site (clothing and spade, described above) on nearby peat pipes shows these natural peatland features as very distinct on GPR profiles.

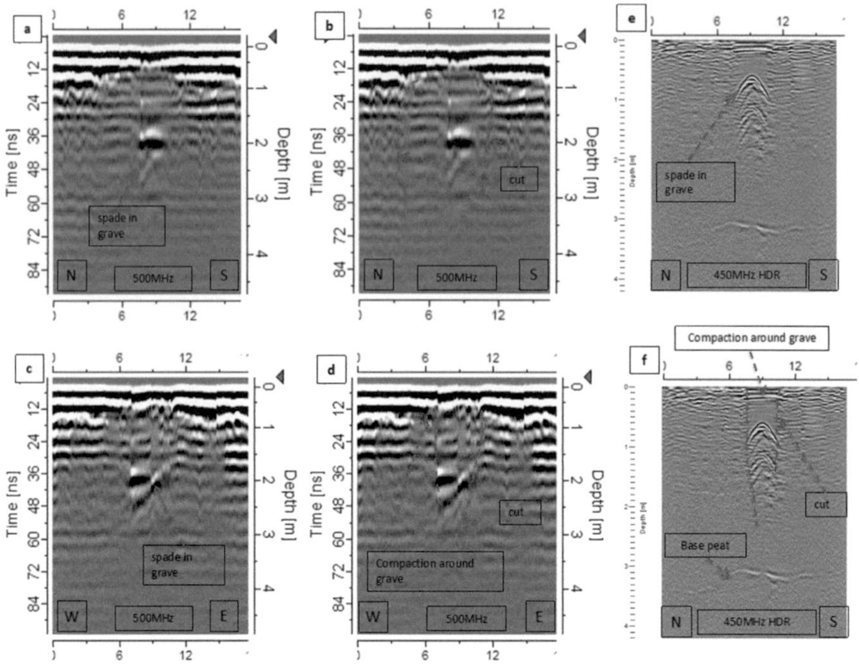

Fig. 4.3 GPR in an upland peat bog in the Pennines, northern England (UK), showing a comparison of an older (manufactured 2006) 500 MHz shielded GPR antenna (Mala Geoscience) and a newer (manufactured 2016) 450 MHz, high dynamic range (HDR) antenna (the latter courtesy of Mike Langton, Mala Geoscience), as in Fig. 4.2. All data are unprocessed. (**a**) 500 MHz data across a test location where a spade was buried some 10 years previously. Data gathered in October 2015. (**b**) Interpretation of the data shown in (**a**). (**c**) 500 MHz data gathered at a right angle to (**a, b**). (**d**) Interpretation of (**c**). (**e**) 450 MHz HDR GPR data across the same spade burial location shown in (**a, b**). Data gathered in November 2016. (**f**) Interpretation of the data shown in (**e**). As in Fig. 4.2, 500 MHz and 450 MHz data were also collected across a simulated dug grave of the same age, with no contained objects and no GPR anomalies (data available on request)

4.5 The Geophysical Detection of Human Tissue in Freshwater

There are excellent review articles on the use of terrestrial forensic geophysics (Pringle et al. 2012). For the search of water bodies, Ruffell et al. (2017) provide a synopsis of the detection of human bodies (and other objects of forensic/environmental crime interest) submerged in freshwater. These authors surveyed a live person in a swimming pool with a floating GPR or water-penetrating radar (WPR). A criticism of this is that the clear contact shown on WPR may be derived from the residual air held in the submerged person's lungs (see above; air pockets in peat or water provide superb radar targets). This is partly vindicated by observations made when involved in police searches for homicide and drowning victims, wherein the boat-deployed radar was manoeuvred over police divers with air lines or

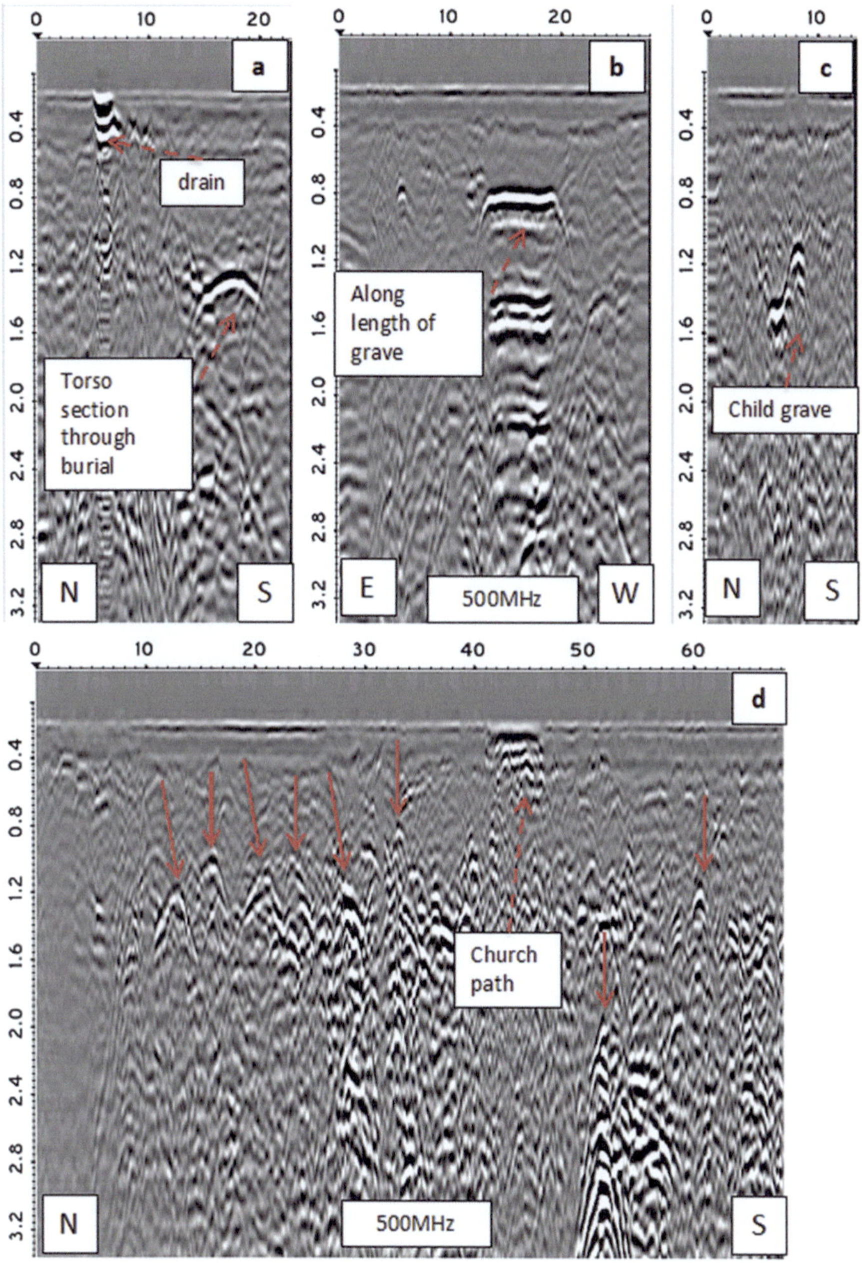

Fig. 4.4 GPR data gathered from a Famine Graveyard (Irish Potato Famine) in a section of a church graveyard in Dowra, County Leitrim, Ireland, where unmarked (no headstone, no wooden crucifix) 'pauper' burials are recorded as 'poor ground' by church authorities. The soil (unknown thickness [from physical probing and 100 MHz regional GPR surveys] but exceeding 11 m) comprises organic-rich lacustrine silts, reworked from adjacent Lough Allen. (**a**) 500 MHz data across a burial, probably around torso location (from the location of a change in vegetation at surface). (**b**) 500 MHz data along the length of an adjacent grave to that shown in (**a**). (**c**) 500 MHz data gathered at a small burial plot (estimated from a slight ground depression), likely an infant or child burial. (**d**) Long GPR section (500 MHz data) through part of the graveyard with supposed burial anomalies marked as arrows

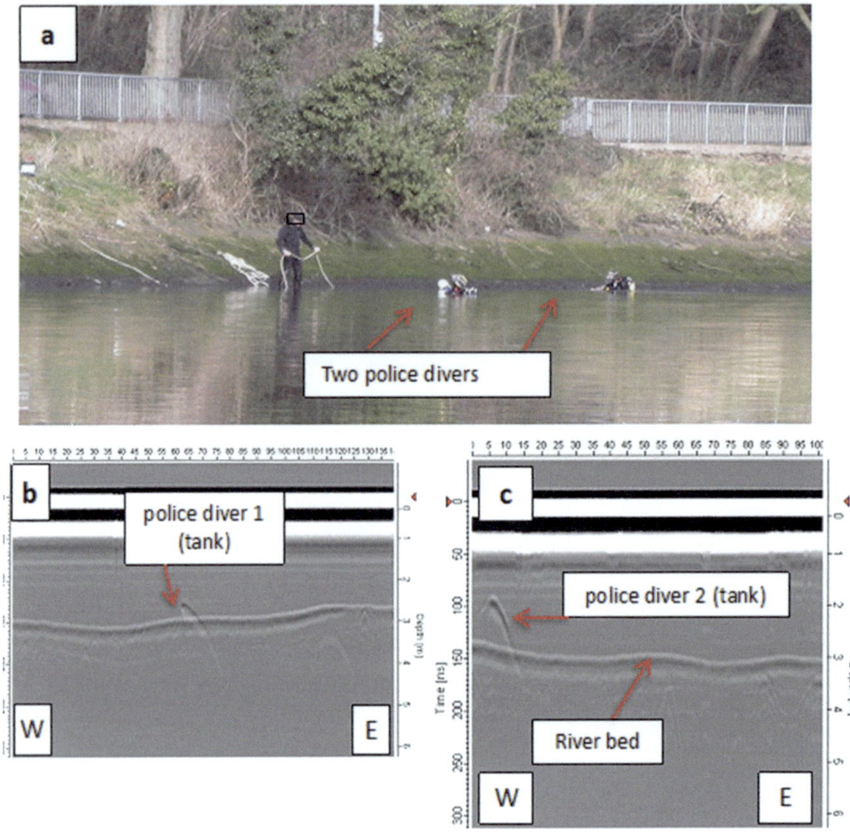

Fig. 4.5 WPR data gathered in the River Lagan, Belfast (N. Ireland, UK), from a 4 m-long inflatable boat with the foot slats removed for maximum antenna coupling to the water: only 4 mm of plastic separate the base of the radar antenna from the top of the water. Data collected as part of a police search for a homicide victim (the body recovered in this location and shown in post-mortem examination by the pathologist to have been murdered 4 months previously). The victim's burntout vehicle was found in a country road lay-by some 10 km upstream of this location. (**a**) The search location, showing the two police divers and the safety officer. Permission to survey the divers obtained. (**b**, **c**) 250 MHz WPR data gathered as the boat was steered over each diver, showing the clear hyperbolae from the position of the diver (visually observed under the boat)

non-metallic air tanks, used in order to deploy waterproof minimum metal-content detectors. Clear hyperbolae were observed over the divers (Fig. 4.5), possibly caused by their air tanks, residual air in their diving suits or (of greatest interest here) the actual human tissue itself, albeit that even this has gas in it (in the blood or from decomposition when deceased).

The question remains: Can human tissue with minimal air/gases be detected by WPR? To answer this question, a scaled experiment was conducted whereby we surveyed real human limbs in plastic and concrete tanks of freshwater at a range of frequencies (1.2, 1.6 and 2.3 GHz), both from above (to simulate WPR from a boat) and from the side (to simulate the search for cadavers in water storage tanks, farm slurry tanks). Water storage tanks are known to have been used as body disposal locations, the best known being the case of Elisa Lam, a student from British Colombia (Canada), whose body was found in the rooftop water tank on the Cecil Hotel, in Los Angeles. The authors have likewise assisted the UK police and search and rescue organisations with searches of farm slurry tanks, concrete water storage reservoirs and beer/cider vats for both victims of homicide and accidental death (usually overcome by fumes). The results of this scaled-down simulation are shown in Fig. 4.6.

Previous to this experiment, we located a sediment-filled concrete tank and surveyed live human legs at 1.6 GHz and 1 GHz, to simulate burial in water-logged sediment. We then surveyed our horizontal live human arm from above by skimming the hand-held antenna over the water surface and our vertical leg by scanning from the side, all with success. The consideration that forensic targets may be in sediment raises a further consideration: Can water-penetrating radar (and other geophysical methods, such as CHIRPS) be used to map sediment thicknesses in water bodies and thus be used as a reconnaissance method? The idea is similar to Harrison and Donnelly's (2008) concept of 'diggability' in terrestrial searches; only here 'sinkability' is considered. Sinkability is suggested to be locations where there is a sufficient thickness of soft sediment in a water body to cover an object. The size of the target will dictate whether sediment thickness could obscure an item from view or sonar: for instance, a small weapon could be covered by a few centimetres of mud/silt/sand, whereas a human body would require some decimetres of sediment cover and a vehicle (depending on size), maybe metres of sediment. The idea here is that should no floating or water-bottom target be found, yet there is a high level of intelligence-based assurance that an object has been hidden in the water body under investigation, then where might search resources be targeted for objects that have sunk into loose and unconsolidated sediment? WPR profiles may not just provide images of water and shallow targets but also sediment thickness (Fig. 4.7a, b). These thicknesses may be digitised or brought into 3D software (Reflex 3D, ArcMap, 3DGeo) as contour plots. Geolocated, such plots can be navigated in real time using a GNSS/GPS or can be redrawn in simple form for non-GIS familiar search teams to use as a way of visualising their search location. The authors have successfully done this for searches of ponds and small lakes; the type of user-friendly output is shown in Fig. 4.8a, b. Such maps are effectively RAG prioritisation diagrams, with access, visibility and sediment thickness promoting high-probability search locations, while visible areas, with thin subaqueous sediment or rock at surface having a low likelihood of containing a sediment-sunken object.

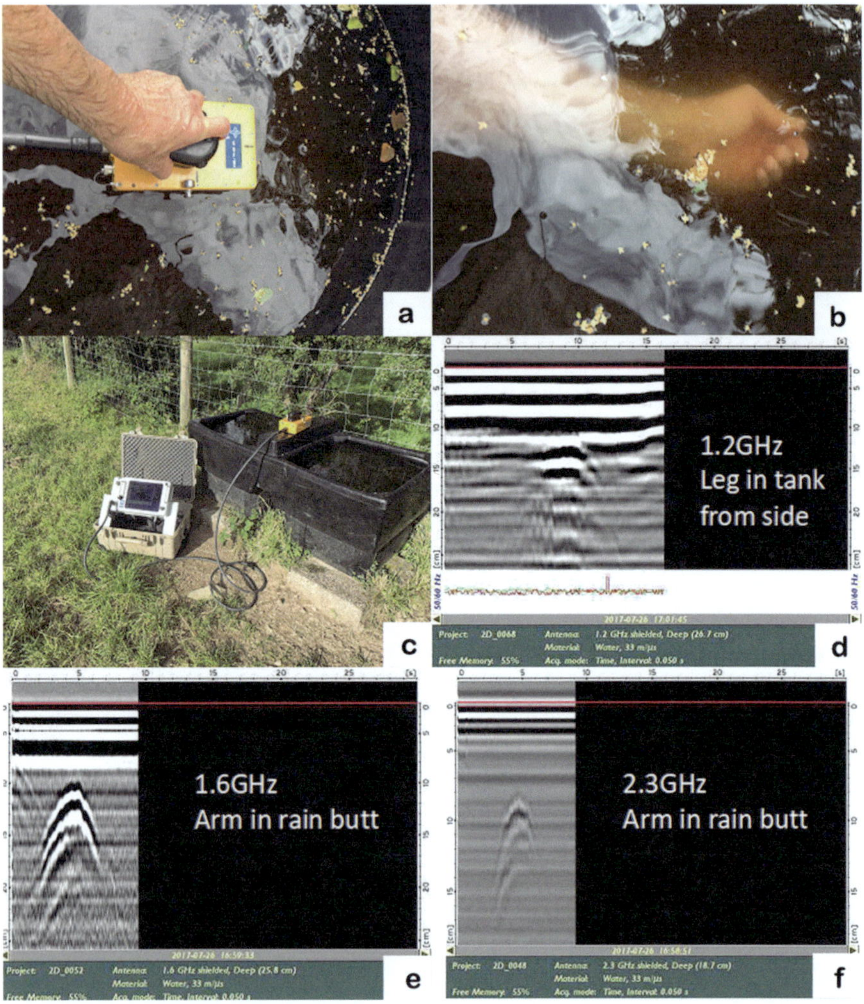

Fig. 4.6 High-frequency WPR data gathered from above and from the side of a plastic rain butt (freshwater rain) and plastic animal drinker (chlorinated drinking water, mixed with rainfall), with live human limbs inserted and the antennas moved across the target(s). (**a**) The rain butt with 1.6 GHz antenna above. (**b**) The live human arm orthogonal below the antenna in (**a**), simulating a boat-borne search. (**c**) The plastic animal drinker. (**d**) 1.2 GHz data gathered (orthogonal) across a live human leg in the animal drinker, to simulate a lower frequency (e.g. 250 or 450 MHz) WPR 'sideways' search of a larger water storage facility, such as a water storage tank (see text for the description of the Elisa Lam case). (**e**) 1.2 GHz GPR data collected on the plastic rain butt over a live human arm. (**f**) 1.6 GHz GPR data collected on the plastic rain butt over a live human arm. (**g**) 2.3 GHz GPR data collected on the plastic rain butt over a live human arm. Other high-frequency WPR experiments and results using different human subjects, targets and locations (e.g. the concrete animal drinker) are available from the authors

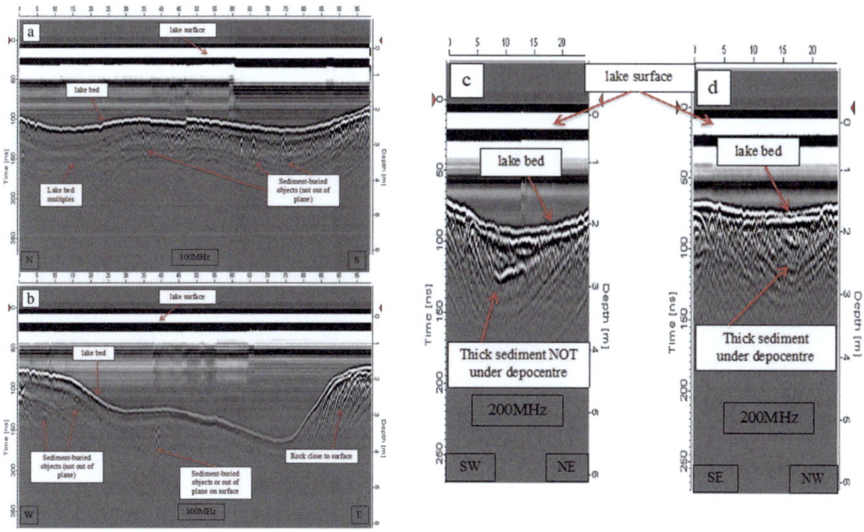

Fig. 4.7 Lake morphology and sediment thickness mapping for search focus. (**a**) Example of a small (100 m diameter) inter-drumlin lake in County Down, N. Ireland (UK), where the centre of the lake is not coincident with the deepest water (which is commonly assumed for searches and coring for palaeoclimate records). (**b**) Example of an asymmetric lake bed, with the radar facies commonly seen in bedrock sub crops (eastern side). (**c**) An example of where the thickest sediment (and thus a likely location for sub-lake bed buried objects) is neither coincident with the deepest part of the lake nor its geographic centre (as in Fig. 4.7a). (**d**) The same WPR survey location as in (**c**), likewise showing the slight offset (in this oblique orientation) between the thickest sediment and the lake depocentre, which are almost coincident

4.6 Conclusions

From experimentation we have replicated and demonstrated criminal casework, stimulated by previous review articles, and provided an overview of how forensic geophysics is being used in nonconventional search locations, such as the inside of engineered structures and on water bodies and peat. This is not to demean the more popular view of forensic geophysics, where buried items are still commonly found in loose deposits under gardens, agricultural fields and unlithified sediments (e.g. floodplains, avalanche slopes, sand dunes). Furthermore, the efficiency of GPR, EM and magnetometry over freshwater and peat suggests that aerial platforms (cf. drones/unmanned aerial vehicles) may soon be deployed for reconnaissance searches of such environments, along with ice and snow (not discussed in this article). More challenging locations, such as mixed clay soils, metal-engineered structures/foundations and abandoned mine workings, also require a desk study and reconnaissance investigations and combination with remote, ground (or water) surveys with other appropriate search methods such as scent dogs, divers and of course eventual excavation and forensic recovery.

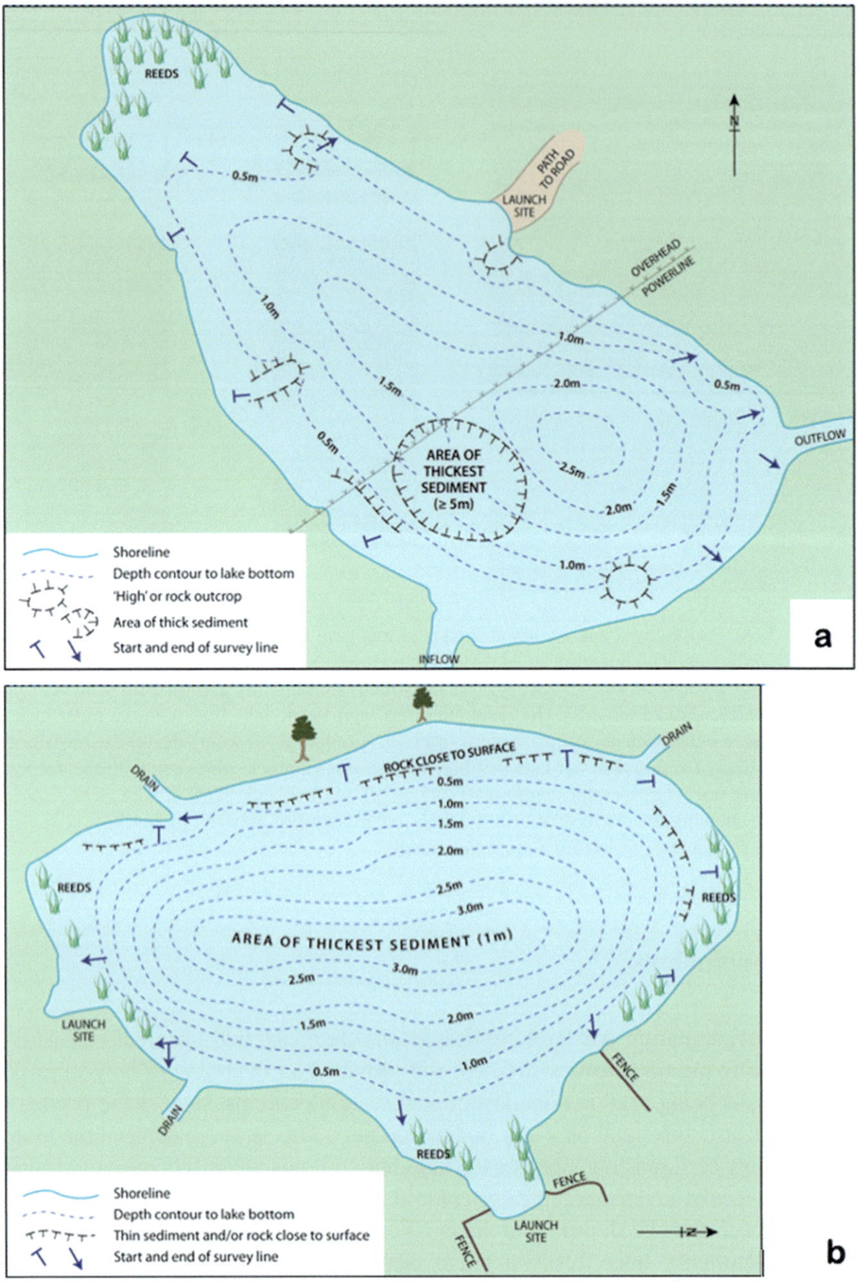

Fig. 4.8 Example of simplified maps of lakes for search personnel. (**a**) Derived from the WPR data shown in Fig. 4.7c, d, showing the dive search team where both the deepest part of the lake is (dashed contours) and the thickest sediment (for buried objects). The mapped length of the overhead powerline (shown) is 42 m, for scale. Such simplified, user-friendly maps are provided for dive personnel to search. The area of thickest sediment ('sinkable area') was probed, using a carbon-fibre probe (50 cm sections) to release sediment gases, allowing boat-deployed scent dogs to be taken over the site. (**b**) Example guide search map of a lake where the deepest water is coincident with the geographic lake centre. The lake is 32 m north to south, for scale

Acknowledgements We are grateful to Mike Langton (Mala Geoscience) for field assistance and advice. Sygma Solutions and Mike Langton facilitated access to high-frequency Mala Geoscience GPR antennas (Figs. 4.1, 4.2 and 4.3). AR wishes to thank Sree Nanukuttan, Geoff Davis, Rachael Parker, Brian Johnston and Lisa Coyle-McLung for access to field sites and invaluable assistance (Fig. 4.1). Sean McAllister (PointPro Ltd) funded the work in Fig. 4.5. Graham Kissock (Police Service of Northern Ireland) is thanked for involving AR in the search shown in Fig. 4.5. Lisa Coyle-McLung and Keith Bennett prompted the work shown in Fig. 4.8. Mark Harrison (Australian Federal Police) and staff of the Centre for Advanced Science and Technology (UK) are thanked for their advice.

Part II
Forensic Archaeology and Human Remains

Chapter 5
The Increasing Role of the Forensic Anthropologist in the Search for the Missing

Nicholas Márquez-Grant

Abstract Forensic anthropology has played a major part in the investigation of human rights' cases and is increasingly playing an important role in searching for missing persons, and assisting in the identification of the deceased in mass fatality incidents. Since its early years, work which was primarily restricted to the laboratory and mortuary setting with the purpose of identification of the deceased, to the present day the role of the forensic anthropologist has expanded to crime scene attendance in order to understand the contextual information in which human remains, primarily skeletonised, have been found and to assist in their recovery. In particular since the 1990s, forensic anthropologists have been key team players in mass grave investigations. Moreover, in recent years, at least in the United Kingdom, there has been an increasing request for forensic anthropologists to assist in search operations for human remains.

This paper provides a brief overview of the role of the forensic anthropologist, focusing on search in particular, and it highlights its value in a number of scenarios relating to searchisng for the missing.

Keywords Forensic anthropology · Human remains · Search · Missing person · United Kingdom

5.1 Forensic Anthropology

Much has been written about forensic anthropology, its definition, its history, the work undertaken, and the methods employed (e.g. Komar and Buikstra 2008; Tersigni-Tarrant and Langley 2017; Dirkmaat and Cabo 2012; Klepinger 2006; İşcan and Steyn 2013). Indeed, forensic anthropology in its wider sense does not only deal with the dead but also with the living (e.g. see Thompson and Black 2006; Cattaneo 2007; Cunha and Cattaneo 2006; Black et al. 2010a; Meadows 2011;

N. Márquez-Grant (✉)
Cranfield Forensic Institute, Cranfield University, Defence Academy of the United Kingdom, Shrivenham, UK
e-mail: n.marquezgrant@cranfield.ac.uk

P. M. Barone, W. J. M. Groen (eds.), *Multidisciplinary Approaches to Forensic Archaeology*, Soil Forensics, https://doi.org/10.1007/978-3-319-94397-8_5

Wilkinson and Rynn 2012; Cummaudo et al. 2014); whether it is to assess the age of living individuals through radiographs and CT scans, or the age subjects from photographs in cases of child pornography, to undertaking gait analysis or analysing physical features from CCTV footage to identify the individual, and even interviewing families in a human rights context. It may also be considered that those anthropologists examining primarily human remains, the body or the skeleton, may be more specifically called forensic physical anthropologists, since other anthropologists – whether social or cultural – may be undertaking forensic work too (e.g. Rosen 1977; Berger et al. 2015; Sanders 2005; Gill et al. 2009). In addition, the relation between forensic anthropology and archaeology and forensic practice overall differs according to country (e.g. see Kranioti and Paine 2011; Márquez-Grant et al. 2012; Groen et al. 2015; Ubelaker 2015). This relationship between archaeology and (physical) anthropology is not a topic for discussion here; although for the purpose of this paper, archaeology is excluded and also covered elsewhere in this volume.

Stewart (1979: ix) defined forensic anthropology as *that branch of physical anthropology which, for forensic purposes, deals with the identification of more or less skeletonised remains known to be, or suspected of being, human.* He further adds (Stewart 1979: ix) that apart from establishing if the remains are human, *the identification process undertakes to provide opinions regarding sex, age, race, stature, and such other characteristics of each individual involved as many lead to his or her recognition.* Forensic anthropology, therefore, can be defined, simplistically and in other words, as the application of methods and principles of physical anthropology[1] to cases of medicolegal or forensic interest. However, humanitarian cases may also fall under the umbrella of forensic anthropology.

Today's definition has also expanded beyond assisting in the identification of the deceased. It involves assisting in the location and recovery of human remains, which are in a certain state of decomposition and cannot be rapidly identified and also addresses trauma, which may have been associated to the manner and cause of death. Komar and Buikstra (2008), following the American Board of Forensic Anthropology state, more specifically, that the discipline deals with the identification of skeletal or badly decomposed or unidentified human remains; assists in the location and recovery of human remains, mainly of unexplained or suspicious deaths; and seeks to provide an opinion on post-mortem interval and evidence regarding any foul play. It was Dirkmaat and colleagues (2008a, b) who highlighted new trends in the field: the increasing role of DNA analysis, forensic archaeology, forensic taphonomy and trauma. It has developed from a role in the laboratory to attending crime scenes and on occasions now certifying the cause of death alongside forensic pathology in some countries or states. Developments in the field (e.g. see Latham et al. 2018) are also shaping these trends. Analysis may be done directly

[1]The American Association of Physical Anthropology (http://physanth.org/, accessed January 2018) defines it as 'a biological science that deals with the adaptations, variability, and evolution of human beings and their living and fossil relatives. Because it studies human biology in the context of human culture and behaviour, physical anthropology is also a social science'.

on the bone or through imaging and remotely. Specialisation is also necessary with this developing trend and the new techniques, and whereas in the past one forensic anthropologist may have analysed a skeleton, it may be that now several specialist forensic anthropologists (e.g. histological analysis, trauma analysis, taphonomy, age at death, craniofacial reconstruction, etc.) work on the same case.

For the purposes of this paper, the role of the forensic anthropologist relates to the dead or those presumed dead. It relates, in particular, to the contribution forensic anthropologists can provide as part of a wider team in the search for the missing. Their role is becoming vital (e.g. see Hackman 2016) in cases of missing person searches, searching for human remains in mass fatality incidents and searching for the right remains during an exhumation; case scenarios which are discussed below. The below-mentioned discussion does not imply that the forensic archaeologist has no role to play in forensic investigations involving search for and recovery of clandestinely hidden human remains. The focus is to inform about the possibilities a forensic anthropologist has to offer, especially during casework in which the forensic archaeologist and police officers have limited knowledge of human osteology and anatomy as well as on forensic taphonomy. Finally, it is recognised that some of the roles mentioned below will be fulfilled by a forensic archaeologist, depending on the country.

5.2 What Questions Can a Forensic Anthropologist Answer?

Although there are a number of forensic (physical) anthropologists working on identifying living subjects via CCTV images or photographs, for example, or on estimating the age of living individuals (Cattaneo 2007; Black et al. 2010a; Wilkinson and Rynn 2012); with specific regard to the dead and particularly skeletal remains, the role of the forensic anthropologist can be summarised as follows (after Komar and Buikstra 2008; Roberts and Márquez-Grant 2012; Márquez-Grant 2015; Márquez-Grant et al. 2016a, b):

- Assisting with search, location, documentation and recovery of either surface, buried or submerged human remains, working together with a forensic archaeologists or using forensic archaeological methods.
- Assessing whether the material is bone.
- If it is bone, whether it is human or non-human.
- Providing an inventory of skeletal elements present, indicating which ones are missing and providing an explanation as to why that may be.
- Commenting on the number of individuals present.
- Answering the question: What information can you obtain from these remains?
- Assisting in the identification of the deceased by reconstructing a biological profile (e.g. age at death, sex, stature, ancestry, unique identifying features).
- Indicating whether or not a skeleton or skeletal remains are consistent with a missing person's biological profile.

- Reconstructing fragmentary remains for a better assessment of trauma.
- Attributing bone fragments or complete skeletal elements to a particular individual or a primary identifier (e.g. dentition) in a commingled assemblage and/or minimising the number of DNA samples to be taken.
- Providing an opinion on the time-since-death or post-mortem interval.
- Interpreting taphonomic modifications to the bone; for example, have they been burnt, cut or bleached?
- Taking samples for DNA analysis, identifying and documenting the bone prior to sampling.
- Assessment of trauma, evidence of neglect or torture.
- Assessing whether the damage on the bone is post-mortem.
- Undertaking craniofacial reconstruction.
- Giving advice when required on other techniques of identification.
- Providing general advice, report writing and attending court as an expert witness when required.

The scenarios in which these questions relate are domestic homicide cases, natural and accidental deaths, mass fatality incidents, genocide cases, etc. The bulk of the work, once the remains have been confirmed as human, is assisting in the identification of the deceased. Whilst biological profile in a laboratory or mortuary setting will assist with the identification of the deceased, for example, by reducing the list of missing people to whom those remains could belong after obtaining information on age at death, sex, stature, etc., this paper focuses on the aspect of searching for the missing (presumed dead) in the field. Indeed, in recent years, at least in the United Kingdom, there has been an increase in forensic anthropology requests by the police for search operations,[2, 3] and an awareness of its importance is increasing through training and through the academic literature (e.g. see Hackman 2016; Márquez-Grant et al. 2016a, b). Therefore, from the above list of questions the forensic anthropologist can address, those in particular relating to search are emphasised below and addressed later in this paper:

- Search and recovery. For example, in areas where there are many bones and bone fragments but non-specialists are unsure if the bone is human or not. Prior to the search, the forensic anthropologist can provide a taphonomic report, contribute to the forensic strategy (e.g. recommended sieve size for human remains) and brief the search teams. The forensic anthropologist can also assist with identifying the right individual where many individuals may be buried in the area (e.g. cemetery, mausolea, etc.).
- Is it bone? Questions may be asked by non-specialists, especially when searching for small bones in cases of missing children, regarding whether material they

[2] http://www.mirror.co.uk/news/world-news/ben-needham-police-to-start-digging-1384886 (accessed January 2018).

[3] http://www.mirror.co.uk/news/world-news/madeleine-mccann-search-april-jones-3642420 (accessed January 2018).

have found is bone or not. These questions may be asked during the walking search phase or during the sieving process for example.

- Is the bone human or non-human? When search officers find a bone, the forensic anthropologist can assess whether it is human or not by direct visual observation or, if remotely, through photographs or other imaging means. If the forensic anthropologist is on site, then a quick assessment can be invaluable.
- What skeletal elements are present? Which ones are missing and why? This is information that is given to the police and can inform search teams regarding what skeletal elements need to be found. An assessment of modifications to bone, such as those as a result of scavenging activity, may explain the missing elements, likelihood of retrieval and with the right specialist, define the wider search boundaries.
- How many individuals? Sometimes the search, which is intelligence led, may recognise that the individual to be searched is to be found alongside a certain number of individuals. This may be asked in mass grave scenarios, exhumations from cemetery niches or vaults, mass fatalities, fire scenes, etc.
- Does this skeleton match the missing person's profile? When searching for the remains of a particular individual in a cemetery (e.g. an exhumation to obtain a DNA sample), the forensic anthropologist will need to take into account as much ante-mortem information as possible about the individual. Although ultimately there will be a primary identifier (e.g. odontology, DNA), in a cemetery with many buried, the assessment of the biological profile in situ is necessary in order to exhume the right individual.
- Any trauma? This may be applicable, for example, in a scenario where the deceased may have been executed in a human rights' case, buried in a normal cemetery, and then a few years later, a request is made to search for the remains for later repatriation and reburial. Apart from the biological profile, the ante-mortem information regarding gunshot trauma, for example, is important in terms of finding the right person (in cases where other people buried in the cemetery were not executed).

Thus, this paper aims to describe and provide an awareness of the valuable role of the forensic anthropologist in the search for the missing (presumed dead) and to introduce the ways in which forensic anthropology can be practised in search operations. This knowledge derives primarily from a personal reflection through experience with forensic casework, supported by the academic literature. It is, mainly, providing a perspective from a forensic practitioner's point of view and as a direct witness in recent years of the increasing role of forensic anthropology in search operations. Although forensic archaeology may come under forensic anthropology in a number of countries, forensic archaeology (see Hunter et al. 2013; Groen et al. 2015; Márquez-Grant et al. 2012) is treated separately here and excluded from this paper, so that focus can be given to forensic anthropology with regard to expertise in the human skeleton, in this case concerning the dead, rather than an assessment of the skeleton in the living. As the objective is forensic anthropology, this paper focuses on searching for human remains where the remains may be skeletonised, bones disarticulated and/or fragmented.

This paper is structured according to a number of hypothetical scenarios yet deriving from the author's experience in forensic and humanitarian casework. Thus it is divided into (a) search of missing persons, presumed dead, for example, in the case of a missing individual and as part of a police investigation; this is the longest section; (b) mass fatality incidents; (c) exhumations; and (d) other scenarios.

5.3 Forensic Anthropology in the Search for the Missing

During a search, the forensic anthropologist will work integrated within a team. His or her tasks can include briefing the search team regarding, for example, the shape and size of human bones and walking alongside or behind the team as they are actively searching the area flagging material of interest and asking for an immediate expert opinion on the following questions: Is it bone? Is it human? This search may be undertaken in a variety of settings, with juvenile remains as a result of homicide found in attics, basements, fields, gardens and quarries amongst other contexts (e.g. see Gill-King 2009).

Whilst search strategies and processes in forensic archaeology may be classified into a number of phases (Hunter et al. 2013), so too can this occur in forensic anthropology. For instance, the equivalent of the archaeological (desk-based/initial) assessment may be in the form, from the forensic anthropological perspective, of taphonomy reports on the human remains; this could be followed by anthropological input in a team briefing or training, after which the presence of the forensic anthropologist is recommended during the active search to assess whether what has been found is a human bone. Finally, a final debrief may be put in place to advice on further work stemming from the search, for example, the need to search for missing skeletal elements, the need to consider sieving the material which has been excavated and retained, etc. Thus, Márquez-Grant and colleagues suggest a number of possible steps or phases from the point of view of forensic anthropology (modified from Márquez-Grant et al. 2016a):

1. Taphonomy report and assist in designing the search strategy (desk-based assessment): Do children remains survive in this type of environment? Will the remains be skeletonised after such a time period? Has there been any ploughing or any human activity on the land since the person went missing? Are there any scavengers, and how may these affect the skeleton?
2. Team briefing on site: What to look out for? What do bones look like in this environment? Will they be fragmented? What size are the remains?
3. During the active search: Is it bone? Is it human? Opinions can be provided remotely or on site; as Komar and Buikstra (2008: 95) point out, the forensic anthropologist can be part of the visual or pedestrian searches. It may be that

endoscopes have to be placed in certain areas and the forensic anthropologist can advise, highlighting the limitations, whether or not human bones can be observed.
4. Assess what is being excavated: Advise whether the remains are from the right person or constitute similar or consistent biological profile from the ante-mortem or historical information available on a particular individual, be involved in the sieving process, and advise on what skeletal elements or parts are missing.

The above first three points are discussed in more detail below.

5.4 Taphonomy Report

Taphonomy reports in forensic anthropology have been recommended elsewhere (e.g. Márquez-Grant et al. 2016a, b). Taphonomy, as we know, relates traditionally to the laws of burial, or the transition from the biosphere to the lithosphere, the conversion from bone to fossil following Efremov (1940b). Haglund and Sorg (1997) define 'forensic taphonomy' as the study of the processes affecting the body after death or post-mortem, whilst Dirkmaat (2012) indicates these processes go beyond the discovery and recovery of the human remains but much further into transportation, laboratory analysis and curation (see also Schotsmans et al. 2017a, b).

Whilst forensic archaeologists will have undertaken a desk-based assessment on the area to be searched (e.g. see Hunter et al. 2013), this information may serve for the report on the human remains written by the forensic anthropologist. As such, it is believed that a report on what is expected to be found and how (e.g. state of preservation) may assist in designing a forensic strategy as well as the anthropological briefing to the search team(s). As is known, many factors can influence the decomposition process, including temperature; humidity; deposition surrounding; soil; presence of scavengers; biological profile (including body mass); disease; manner and cause of death; plants; grave attributes including depth, clothing and other wrappings; geological features; etc. (e.g. see Haglund and Sorg 1997, 2001; Pokines and Symes 2014; Schotsmans et al. 2017a, b; Junkins and Carter 2017; Barker et al. 2017). It may be that a combined report by archaeologists and anthropologists would be the most appropriate.

Often, apart from any archaeological desk-based assessment that may be available (e.g. see Hunter et al. 2013), forensic taphonomy reports can be of use as it answers questions specific to human remains that the police want to know prior to searching or planning a search.[4] For example:

- Do the remains survive in this environment?
- Will the remains be fleshed or skeletonised?

[4] I undertake an exercise when training police forces and forensic scientists generally, which is to think of a hypothetical scenario and do a little research to explain to the rest of the class how the body would be preserved in such an environment after a certain period of time, what search strategies are planned for this type of scenario, etc., an exercise which seems to be of value.

- What size will they be?
- What colour will the bones be?
- Will they be fragmented?
- Will the remains be buried or scattered?

To cover what is expected and to relay this information to the police and other specialists, the taphonomy report or initial assessment report may be written with the following structure (see also Márquez-Grant 2015):

- *Introduction*: Hypothesis of what may be expected, deposition of remains and intelligence information.
- *Background*: When did the person disappear? What clothing was he or she wearing? What was the sex and age of the person? Are there any unique identifying features? Did he or she suffer from any disease?
- *Sequence and timing of decomposition of a human cadaver*: What is the sequence of decomposition? Is mummification typical in this environment? What if the body has been buried? What is the geology of the area to be searched?
- *The human skeleton*: If juvenile, then provide information on the size of the bones for a particular age, whether epiphyses are unfused, different types of bones, etc.
- *The search area*: Environmental assessment of possible deposition site – climate, topography, vegetation, landscape use, fauna, depth of certain strata, etc.
- *Assessment*: If the remains would survive, whether they would be scattered in the surface, whether they bones would be stained a particular colour, etc.

The above assessment may lead to a number of bullet points that may be helpful for the police prior and during a search. Below are some summary bullet points from a hypothetical case, and it highlights its implications for search:

- Burial may have occurred by understanding the depth of the soil in the area.
- The corpse with this x amount of time would be expected to be skeletonised in this environment.
- Bones may be fragmented, weathered and incomplete due to the action of the soil and the scavengers in the area.
- Bone fragments on the surface is possible, for example, if there has been ploughing in the land to be searched, which may have disturbed any shallow graves.
- Bones may be stained a specific colour due to this geology, or bleached if exposed to the sun.
- Bones will survive in this environment for hundreds of years.

The above relates to the body, although of course other evidence types can be taken into consideration, such as the survivability of clothing the missing person was last seen wearing (see Janaway 2008; Stuart and Ueland 2017a, b).

5.5 Team Briefing

As the search will be undertaken by a number of individuals, not always trained in human osteology, a briefing by the anthropologist is recommended (Márquez-Grant 2015: 311; Márquez-Grant et al. 2016a, b). Photographs of adult/juvenile bones, bone casts and other materials may be brought to the briefing. The briefing should cover some of the following points (after Márquez-Grant et al. 2016b):

1. Depending on information regarding the biological profile of the missing persons (e.g. infant), provide brief training regarding the appearance and dimensions of the bones and the different types of bone shapes in the skeleton (e.g. long bones, flat bones, irregular bones). This awareness can be provided through photographs, casts, etc. Be aware to mention teeth and dental fragments if a forensic odontologist is not present at the briefing and what developing teeth look like, etc.
2. Based on the information contained in the taphonomic report, inform the team about what is expected regarding degree of decomposition or skeletonisation and the different possibilities that the remains may be disarticulated, weathered, burnt, bleached, fragmented, etc. Even photographs of archaeological bone may show how modified bone can be with the passing of time.
3. Based on the taphonomy report, also indicate the likelihood that the remains may be scattered, brought to the surface and fragmented following ploughing of the land, any scavengers that may have modified and transported the remains to a certain distance, etc.
4. Perhaps it may be necessary to show examples of bone staining, colour of bones, etc. in the area. This may be done if there are a number of non-human bones in the area that can be observed and showing colour changes through sunlight, etc.
5. Provide information about the sieving process, when this will be required and whether manual sieving, automated saving or raking (in some circumstances when there are large quantities of soil to deal with) may be required.
6. Reiterate that the search teams should always consult with the forensic anthropologist and always check if in doubt about something being bone or human. The forensic anthropologist will preferably be present at the scene or during the search operation.
7. Finally, it may be worth discussing during this briefing, the protocol for any documentation, even of non-human bones, with the scientific police or CSI team, the collection of non-human bone during the line search, whether bones that are not easily diagnosed are exhibited, etc.

5.6 During the Physical Search for Human Remains

It is not the aim here to provide an analysis or recommend different search strategies and techniques (e.g. see Hunter et al. 2013); however if excavation with a mechanical digger is required to strip the area in search for a clandestine grave, for example, depending on the intelligence information and taphonomy report, the forensic anthropologist, in conjunction with the forensic archaeologist, may suggest searching the ground for any bones that may have come to the surface (e.g. through plant activity, agricultural use, ground alterations due to severe weather) prior to excavating an area.

From personal casework experience, a line search where the forensic anthropologist is behind that line providing support (Is it bone? Is it human?) is effective. It avoids forensically documenting a bone, which may be non-human, with a rapid assessment and provides time-saving efforts. It allows the search to continue until human remains are found.

If the forensic anthropologist is not present during the search operation, photographs of bones (including a scale) taken from different angles may also be sufficient; however in areas where the grounds have large quantities of animal bones, it may be more cost-effective and necessary to have a forensic anthropologist on site. The forensic anthropologist should and will also be familiar with other tissues, such as calcified tissue or ossified cartilage and any bone anomalies or variations (e.g. bone ossicles), in order to fully assess whether or not a bone or material is human.

Hunter et al. (2013: 200) indicate that sieving provides a second opportunity to recover materials and achieve as much as possible a full skeletal recovery. During the search operations therefore (following Márquez-Grant et al. 2016b), the forensic anthropologist should also be sieving or checking material that has been sieved and retained at sieving stations; likewise if large quantities of soil have been raked. He or she should also provide information on the most appropriate sieve sizes (e.g. 10 mm, 5 mm, 2 mm mesh sizes or variations from that). This advice can be assessed on a case-by-case basis and depends on a number of factors including soil type, material to be sieved, size of missing elements, quantities to be sifted, time constraints, resources and health and safety. Although usually a quick visual assessment on site is sufficient (e.g. see Mulhern 2016), it may be that with such small fragment sizes, a second opinion or further analysis in a laboratory such as histology and DNA is required (see Ubelaker et al. 2004; Ubelaker 2018; Cattaneo et al. 2009; Crowder et al. 2018).

Even if the bones found are not human, it is advisable that the remains are retained and disposed of elsewhere. If human remains are found, the forensic anthropologist, other specialists (e.g. forensic archaeologist), the forensic pathologist (if required and depending on country legislation and type of case) and police teams can proceed with their appropriate forensic protocols. An inventory can be provided by the forensic anthropologist of the bones present/recovered and a list of missing elements can be forwarded onto the police officer in charge of the investigation and/or search teams.

To summarise, in these scenarios the forensic anthropologist should:

1. Brief the team about 'what to look for' – size of bones, types of bones, completeness and potential condition of the remains. Part of this will be based on the taphonomy assessment relating to the human remains, which has taken into consideration time since disappearance, circumstances of disappearance, topography of the area to be searched, etc.
2. Be able to quickly assess whether or not a bone or bone fragment is human, unless it is too fragmented and compromised that laboratory analyses are required. The presence of a forensic anthropologist in these search operations will save time and money by immediately excluding the bones as human (in the case animal remains are found) without the need for further forensic procedures, exhibiting, etc. to take place. This assessment may on occasions be undertaken remotely through photographs for a small quantity of well-preserved bones.
3. Be able to assist and provide advice on sieving procedures and check material which has been sieved.
4. Recommend that non-human bones be retained and documented quickly, after which these will be disposed in alternative location.
5. If the remains are of human origin, assist with documentation, recovery, etc.
6. Have in place a contingency plan; for example, another forensic anthropologist on standby to provide a second opinion – even if remotely via photographs – in cases where identification may be challenging. Also consider other techniques and have a list of who may be able to undertake these analyses.

The high responsibility of identifying bone as human or not human cannot be emphasised enough.

5.7 Mass Fatality Incidents

In a mass fatality incident, forensic anthropologists can be part of a team working to identify the remains, provide answers and bring closure to relatives. Increasingly, forensic anthropologists play an important role in mass fatality deployments and on many occasions lead a number of tasks. Many countries will have a list of forensic anthropologists as part of their Disaster Victim Identification (DVI) teams.

Focusing on the search aspect for this paper, depending on the nature of the incident, fragmentation of skeletal remains may be present, the remains may be compromised (e.g. by fire), and there may be commingling between remains of various deceased and mixing of human remains with animal remains. In these circumstances, the forensic anthropologist's question as to whether a bone or fragment of bone is human or not is crucial in saving time and reducing costs, minimising any DNA analysis. This triage needs to occur at the very beginning (Sledzik et al. 2009: 291; Black et al. 2010b: 345), and it is often led by forensic anthropologists, for example, in the case of the World Trade Center (Mundorff 2014: 375).

Moreover, Sledzik and Mundorff (2016: 478–479) also highlight the value of forensic anthropology, in particular during a mass fatality investigation, to (a) devise an appropriate system for search and recovery of remains; (b) assist with locating, recognising or identifying and recovering human remains; and (c) train or brief search personnel to recognise human remains. This is often done alongside a forensic archaeologist. The benefit of a forensic anthropologist at the site/scene, they argue, is that they can provide an identification of remains even if they have been affected considerably by taphonomic factors, remains that may be overlooked by non-specialists, apart from providing an opinion as to whether the material is human, non-human or non-biological (Sledzik and Mundorff 2016: 479).

In summary, with regard to searching for human remains following a mass disaster, the forensic anthropologist can be part of an initial triage at a sieving station to assess whether bone or bone fragments are human or not. Alternatively, the forensic anthropologist can examine in the laboratory or morgue the material classified as bone or possible human bone by the recovery teams. This system provides a more cost-effective way rather than sending samples directly to a laboratory for DNA analysis. The forensic anthropologist may also be able to comment on minimum number of individuals, which should eventually feed into the search teams. For example, police information may relate to a family in a house that has exploded, but the remains of a certain person do not appear to have been recovered from the biological assessment of the remains by the forensic anthropologist and forensic pathologist at the mortuary. Finally, to minimise DNA sampling and providing as many remains as possible to the families of the deceased, the forensic anthropologist in a commingled assemblage can attempt to attribute as many bones or bone fragments as possibly to a particular individual, via systems of pair matching, biological profiling, reconstruction, etc. (see Adams and Byrd 2014).

5.8 Exhumations

There are a number of reasons for exhuming or disinterring human remains. On some occasions, there is a need to exhume an unknown individual in order to obtain a sample for DNA analysis when after further investigations an identity is presumed or in cases of mistaken identification in the past. On other occasions, exhumation of a known individual is required to review the initial autopsy or because a criminal investigation has been opened. There are many examples of exhumations worldwide due to some of the reasons just mentioned.[5] Other reasons include family

[5] E.g. the exhumation of President Salvador Allende from Chile (http://www.nytimes.com/2011/05/24/world/americas/24chile.html – accessed January 2018); Poet Pablo Neruda (https://www.theguardian.com/books/2015/jun/05/pablo-neruda-poisoning-doubts-fuelled-by-new-forensic-tests – accessed January 2018); the identification of Argentinian soldiers buried in the Falkland Islands/Malvinas (https://uk.reuters.com/article/uk-argentina-britain-falklands/scientists-aim-to-identify-remains-of-argentine-soldiers-on-falklands-idUKKBN18S61Y – accessed January 2018).

requests or public health reasons. Thus, exhumations are also necessary for repatriation or because there is a need for reburial (Ferllini 2002: 63; see also, for example, Brooks and Brooks 1984). Recommendations have included the participation of forensic anthropologists and archaeologists. For instance, with regard to search and exhumation of neonatal and infant remains or 'bebés robados' ('stolen babies') from the second half of the twentieth century in Spain, the documentation recommends the presence of archaeologists and anthropologists during the exhumations[6] (see Cubero 2013). Whilst Hunter and colleagues (2013: 136) indicate that in exhumation activities the forensic archaeologist is attempting to reconstruct the events which occurred and the various life actions, the forensic anthropologist can ensure that the individual exhumed has a biological profile consistent with the person to be exhumed or missing. Indeed, a number of exhumations have been carried out with the presence of forensic anthropologists in cemeteries or areas with clandestine graves in order to find the right body.[7] It is this search aspect that is the focus here.

Search operations within a burial ground or cemetery may involve exhuming an unidentified individual buried in the past in a coffin but within a communal grave with many others; it may also involve searching for the right individual from a number of unmarked graves or identifying the right niche or vault containing the remains that need to be investigated. The forensic anthropologist will rely on ante-mortem information obtained by an investigation team from a number of sources or by the forensic anthropologist himself/herself on interviewing witnesses (in order to ask the right questions). This information about the deceased will target age, sex, stature, ancestry, past medical history and any skeletal abnormalities, any other unique features and cause of death and circumstances around death if known. Thus with this information, it is possible to discriminate some individuals over others (adult vs juvenile; young adult vs old adult).

To summarise, exhuming the right individual in a cemetery by assessing biological profile and other information, such as peri-mortem trauma, is another benefit provided by employing a forensic anthropologist in the search, whether it relates to searching for a particular individual in a cemetery or ensuring the right person is exhumed from a communal grave.

5.9 Other Scenarios

There are a number of scenarios, where there has been a discovery of a surface deposition of human remains or a clandestine grave that can be considered a little further. Thus, even when the remains have been found or have been discovered

[6] https://politica.elpais.com/politica/2017/10/20/actualidad/1508500176_449374.html (accessed January 2018).

[7] For example, searching for the remains of a British hostage found in Lebanon (https://www.telegraph.co.uk/news/worldnews/middleeast/lebanon/6592000/Body-found-in-Lebanon-thought-to-be-missing-Alec-Collett.html – accessed January 2018).

(outside of a search operation for a particular missing person), the forensic anthropologist that attends the scene is also able to offer, after documenting and recovering the remains, an assessment of missing bones or parts of bones. Indeed, the missing elements in a scattered surface scene may be assessed through any scavenging marks present on the bones. It is recommended that the forensic anthropologist undertakes, where possible, an inventory of the remains prior to the transportation of the body to the morgue. Nevertheless, after the forensic anthropological examination in the mortuary, a briefing will tend to be held and this, as well as the expert witness report, should include a list of missing bones. That way, the police can feed this back to the search teams to be deployed if necessary. It may be that the scene is still cordoned and guarded or that a wider search is undertaken depending on the results of the post-mortem forensic pathology and forensic anthropological examination and any recommendations thereafter. By contrast, it may be that it is unlikely that any further investment of time may yield any missing elements. That is, what bones are missing? Why are they missing? What is the likelihood of finding the missing bones? These first two questions should be answered in the expert statement. In addition, sieving may be recommended for any small bones and teeth that may be missing. Evidence of disarticulation, scavenging or dismemberment will also provide further information that will eventually be fed into the search team.

Finally, of course, in cases where identification needs to be carried out, the biological profile obtained from the analysis of the human skeletal remains will result in a search for who that person may be. When searching for potential missing persons to whom the remains may belong, the biological profile will narrow down the list of missing persons. As Rogers (2009) exemplifies for British Columbia, in October 1999, there were 1755 reported missing people; 444 (25%) were female, of those 143 were in the range of 20–40 years and, of those, 38 were non-white/Caucasoid. Therefore, the biological profile alongside other information the police may have can narrow the list of potential missing persons. This biological profile can be publicised, alongside craniofacial reconstruction when undertaken, in a number of missing person websites (such as that of the UK Missing Persons Bureau).[8]

5.10 Concluding Remarks

This paper has attempted to demonstrate the value of forensic anthropology in search, without addressing the area of forensic archaeology, which is covered elsewhere (e.g. see Hunter et al. 2013). Deriving from a number of scenarios, the role of the forensic anthropologist can be of utmost importance in the search for human remains or missing persons presumed or hypothetically deceased. Work on those cases can be for judicial or medicolegal purpose or purely humanitarian in nature. Mass disaster scenarios have also seen an increasing involvement of forensic anthropology, particularly in the triage stages. As Hackman (2016) indicates, the ability to

[8] http://missingpersons.police.uk/en-gb/home (accessed January 2018).

identify potential bone material, recognise and identify human bone at the scene, especially if there is fragmentation, colour changes and other taphonomic modifications, is of vital importance, and the forensic anthropologists can be invaluable here.

Overall, this paper focused on search outlines below, indicating the tasks the forensic anthropologists can contribute to:

- Assist in the search for human remains (mainly if skeletonised remains are expected).
- Write a taphonomy report.
- Contribute to the search strategy (e.g. retain animal bone, where to get a second opinion, etc.).
- Brief the teams as to what to potentially expect and what to look for regarding remains of an adult/child.
- Assist the search teams in confirming if what they have found is bone (or tooth) or not (either on site, at sieving station or via photographs if possible).
- Identify quickly and by visual means whether any bones found are human or not.
- Provide assistance or lead the triage regarding human vs non-human bone in a mass disaster incident.
- Assist with the recovery of fragmented human remains.
- Answer the following if the bones are human: How many bones are present? Have some been scavenged? What bones are missing? How is the body positioned?
- To answer the question of how many individuals are present.
- Assist archaeologists in the excavation/recovery of human remains.
- Advice on packaging and transportation of human remains.
- Supervise sieving.
- Exclude individuals by assessing the biological profile.
- Recover the right individual in an exhumation.
- Provide a list of missing elements after human remains have been discovered, to feed into the search teams.
- Obtain or guide questions when obtaining ante-mortem data.

Amongst the benefits, these can be summarised below:

- Cost-effective and more economical: having a forensic anthropologist at the scene/site can save further unnecessary work in the laboratory.
- Time: by having a forensic anthropologist at the scene/site, it is easy to exclude bones as being animal without having to photograph bones, send them for further consultancy, package them and provide an evidence number.
- Eliminate certain areas, and provide degree of confidence of an area that has been searched for human remains.
- Reduce DNA sampling.

Chapter 6
Forensic Archaeology: State of the Art of Archaeological Techniques in France and Other Latest Developments. A Case Studies in Toulouse Region (South West of France)

Patrice Georges, Christelle Buton, and Éric Crépin

Abstract This paper provides a brief overview of the role of forensic archaeology in France, focusing on search in particular, observing and evaluating different scenarios relating to searching for the missing. These observations, and especially the research, will be focused on the Toulouse region.

Keywords Forensic archaeology · Forensic anthropology · French National Gendarmerie · Buried body detection · GPR · Human remains searching dogs · Metal detectors · Earth-moving machines

6.1 Introduction

Due to its territorial jurisdiction, the institution of French National Gendarmerie is most confronted with cases of missing persons who are believed to have died in a criminal context and who have been illegally buried. Each year, depending on the progress of investigations, research operations are triggered. Despite considerable efforts in recent years, these search operations generally do not utilize all the skills

P. Georges (✉)
French National Institute for Preventive Archaeological Research (Institut national de recherches archéologiques preventives; Inrap), Paris, France

Joint Research Unit TRACES ("Archaeological Works and Research on Cultures, Spaces and Societies"; UMR 5608), Toulouse University – Jean Jaurès (UT2J), Toulouse, France
e-mail: patrice.georges@inrap.fr

C. Buton
French National Gendarmerie, Toulouse, France

É. Crépin
Paratroopers Forces (French Ministry of the Armed Forces), Paris, France

P. M. Barone, W. J. M. Groen (eds.), *Multidisciplinary Approaches to Forensic Archaeology*, Soil Forensics, https://doi.org/10.1007/978-3-319-94397-8_6

available in this field. Although the Forensic Sciences Institute of the French National Gendarmerie (*Institut de recherche criminelle de la Gendarmerie nationale* – IRCGN) recognizes the interest of forensic archaeology (Ducrettet et al. 2013; Schuliar et al. 2015) and has begun to take interest in some of its problems, there is, as yet, no national protocol in France, similar to that in Belgium (*Necrosearch*), for example (Schotsmans et al. forthcoming). There is no single point of contact in the French National Gendarmerie for such type of problems. To date, three specialized services of this force have been called upon. Two are located within the Forensic Sciences Institute of the French National Gendarmerie in Pontoise and another, with which they are not in contact, is located within the French National Gendarmerie Dog Training Center (*Centre national d'instruction cynophile de la gendarmerie* – CNICG) in Gramat (Lot). Moreover, the existence of forensic archaeology in forensic criminal investigations is not yet fully formalized, even though the community of forensic doctors fully understands the issues at stake, and anthropologist intervention in the field is a recognized added value in various cases. Finally, forensic archaeology, which some people prefer to call forensic archaeology for these reasons, is confined exclusively to the research and excavation of skeletons. All the research concerning the caches of buried weapons, buried bags containing jewelry, gold, etc. is within the scope of forensic archaeology. But, at present, there is only one case where archaeological methods have been applied at a crime scene that did not contain human bone.

This observation, and especially the research conducted in the Toulouse region, has recently led to new reflections on the subject, in particular by accompanying new actors whose skills have been put to good use to complement the methods generally used.

6.2 Problematics

Except in one case (cf. *infra*), forensic archaeology has often been used for the search and excavation of buried bodies. This is why the Forensic Sciences Institute of the French National Gendarmerie, the only institution to have tried to formalize it, has integrated it into the department dealing with anthropology. However, as is taught at the university and in the military, in many other cases which are regularly the subject of legal proceedings, the archaeologist could be asked to find and especially document the digging in the ground: drums of buried toxic products, caches of active or non-active weapons, illegal items hoarded by criminals and hidden in the ground, etc. All these cases, which obviously require extensive observation of anomalies in the soil and at the edges of the digging pit, do not integrate the skills of the archaeologist. This is all the more damaging because, in some cases, the use of GPR has required a preliminary scraping. This operation would have made it possible to have a reading of anomalies in the soil and the signs of any digging.

To date, we have only one example in France of the use of archaeology in forensic investigations that has nothing to do with the search for a cadaver. These are the investigations carried out in the context of a bank robbery tunnel in the Toulouse region. The archaeologist employed, whose mission was to find traces of digging, also had the task of managing this site for forensic purposes. It was an interface between the civil engineering specialists and those of the Force, particularly with regard to the earth-moving machine and how to get them involved.

Detecting a buried body requires the intervention of specific resources and specialists who often ignore the operational capabilities of the various actors deployed on the site. In France, Crime Scene Technicians (*Technicien en Identification Criminelle* – TIC) and Criminal Investigation Units (*Cellule d'Investigation Criminelle* – CIC), a special unit with Crime Scene Technicians (*Technicien en Identification Criminelle* – TIC), located in each departmental capital (in the gendarmerie group), and the Crime Scene Engineers (*Coordinateurs des Opérations de Criminalistique* – COCrim) can take help of the following skills and services:

- Ground-penetrating radar located in Sound & Image Recovery Department of the Forensic Sciences Institute of the French National Gendarmerie (*Dpt Signal-Image-Parole*; IRCGN) in Pontoise (Val-Oise) (Fig. 6.1).
- The human remains searching dogs (RRH for "*Recherche de Restes Humains*"), known as cadaver search dogs led by dog handlers of the Dog Unit National Investigation Group (*Groupe National d'Investigation Cynophile; GNIC*), based in Gramat (Lot), in the southwest of France (Fig. 6.2).
- The archaeo-thanatological skills that are taught at the Forensic Sciences Institute of the French National Gendarmerie (Coulombeix et al. 2014) and in some

Fig. 6.1 Ground-penetrating of the Forensic Sciences Institute of the French National Gendarmerie during an experiment

Fig. 6.2 Cadaver dog in action (real case)

French forensic laboratories but which are not used as they should be. But in these laboratories, anthropologists are not involved in the search of bodies, only sometimes in discoveries. In cases concerning the Toulouse region, they were carried out by an agent of the French National Institute for Preventive Archaeological Research, trained at the Forensic Sciences (Paris 5 University) (Georges 2009) in addition to his degrees in physical anthropology and archaeology. He is also a forensic expert in the field of anthropology (crime scene and identification according to the nomenclature of French experts). Although it is not systematically implemented, it is increasingly recognized that this archaeo(thanato)logical knowledge is absolutely essential in such crime scenes. Forensic doctors, who are not trained for the excavations, are the first to recognize it (Baccino 2011). The forensic archaeologist is indeed able to interpret the signs of digging in the ground, often invisible on the surface, but which appear when scraping is carried out. This is why it is recommended that a suitable mechanical machine be used (Georges et al. 2012, 2013) (Fig. 6.3). In addition to their expertise, the intervention of archaeologists at crime scenes is also explained by the fact that archaeologists have, in recent years, come closer to more recent chronological contexts and are being recognized for their ability to manage large-scale archaeological sites (Georges 2009, 2016).

There is no obligation for the magistrate or investigator to summon a specialist. There is no protocol on this, which is why the employment rules that we have begun

Fig. 6.3 Archaeological scraping (real case)

to implement through university courses and during the training of the French National Gendarmerie are not yet widely applied, including by the Forensic Sciences Institute of the French National Gendarmerie, which is the driving force in this field.

For organizational reasons, the ground-penetrating radar team is more often involved in the search for bodies, without archaeologists and anthropologists accompanying them. These signal specialists do not have any knowledge of digging, taphonomy, etc., and especially of the ground-penetrating radar, which has enormous limits in rough terrain, the forest (cf. *infra*), and clayey and humid soils. It appears to the people who solicit it (and who do not know its operating principles) as the best tool to find the victim. In these times of technophilia, the presence of a screen allowing a so-called direct reading makes it easy to explain it. This is why people don't really understand that the absence of a result with only the ground-penetrating radar alone is not a result. Numerous cases, some of them highly publicized, are concerned by this methodological flaw in France. It should be noted, however, that many other geophysical methods could be used, but generally require post-processing of the data (no direct reading). This problem of the expression of the result is similar to that with dogs dedicated to the search of human remains with a 6-year-old record in France. While older contexts may be favorable, the absence of results for older cases is irrelevant, including for more recent cases, knowing that it is recommended to create wells of odor (Fig. 6.4) close together (at least after every meter) within a well-defined period of time, which is not always possible. In fact, as Anglo-Saxon literature shows, only archaeological investigations, which include scraping with earth-moving machines, give the assurance of a result. Of course, this poses a problem for open areas, large surfaces and inhabited house plots with landscaped gardens.

Fig. 6.4 Doing smell wheel (real case)

6.3 Observations

In the Toulouse region (Southwestern France), recent criminal cases have led investigators to suspect, several years after the commission of the facts, that the victim was illegally buried in a wooded area where it has not always been possible to restrict the investigation to less than a few hectares. In this environment, it's easy to hide the body using the immediate camouflage of the stigmas resulting from the digging of a pit (presence of a permanent carpet of leaves), whose anomaly of relief tends to disappear in the passing years. It was not possible to engage the ground-penetrating radar in such cases. The strong root development of trees prohibits the use of ground-penetrating radar. The experts of the Forensic Sciences Institute of the French National Gendarmerie say that the numerous roots would create many anomalies, making it impossible to distinguish them from a buried body signal. The length of time since the alleged illegal burial and the environmental context are sufficient to explain why we consider this as a particularly difficult operation. Moreover, the fact that in each case the research area extends over several hectares confronted the survey actors with a new difficulty. This is the reason why it was decided to hire new specialists for this type of operation: the Military Search (*Fouille Opérationnelle Spécialiée; FOS*) of the 17th Parachute Engineer Regiment (*17ème Régiment du Génie Parachutiste; RGP*), which had already been called upon in the region to search for firearms and ballistic elements. The 17th Parachute Engineer Regiment is a French military unit based in Montauban (Tarn-et-Garonne). It is organized, equipped, and trained to carry out the engineering support missions of the 11th Parachute Brigade, for which it provides all the specific tasks of the assault engineer in an airborne, helicopter-borne, and mechanized environment, such as assault clearance, depth reconnaissance, deployment assistance, mine clearance, and clearance operations (munitions, explosive devices, etc.). It has had a constant presence since 1975 in all the conflicts in which France was involved.[1]

The Military Search team of the 17th Parachute Engineer Regiment was born in 2008. It was the first of all Military Engineer Regiments to be created and engaged in external operations. The parachutists who make it up are in charge of implementing technical and scientific forensic techniques for military purposes for gathering information (Lafaye 2012). The primary mission of the Military Search team is to combat improvised explosive devices (IEDs) by collecting clues, in the same way as Crime Scene Technicians, to stop the actors of an attack. In external operations and in front of combat troops, this results in the search for explosive or ammunition caches and possibly DNA samples and fingerprints. In order to retrieve intelligence from the field where it is located (housing, vehicles of all kinds, personnel, etc.), these soldiers have special equipment (Fig. 6.5). The metal detectors in their possession, some of which are specific to this section of specialists, is one of these (see below) (Fig. 6.6). Accustomed to working in various environments to highlight points of interest (landscape anomalies) and anomalies of all kinds and walking long distances and settling in degraded environments, these soldiers proved to be an essential accompaniment to the usual arrangements. All the more so because their technological means of self-staffing add a skill that is not found anywhere else, namely, the detection of metallic element present on and/or in the body, which had never been done before, at least in such a systematic way. However, in body search operations, the detection of metallic elements present on telephone, jewelry, etc. or in the body (ballistic or medical elements) had never been used. Indeed, the involvement of this Military Search team in the national territory for the benefit of the National Gendarmerie was not an aberration, even if its missions during external operations are different. In fact, as engineer soldiers, they were particularly receptive to the discourse on the use of earth-moving machine in such situations; they can drive it (Fig. 6.7). These soldiers have adapted and refined its procedures, which were previously designed to work in a very limited time slot and in a hostile environment (Lacrêpe et al. forthcoming).

[1] http://www.defense.gouv.fr/terre/l-armee-de-terre/le-niveau-divisionnaire/3e-division/17e-regiment-du-genie-parachutiste

Fig. 6.5 Military and forensic equipment of the Military Search team

Fig. 6.6 Metal detector operator in action during an experiment

Fig. 6.7 Scraping by a soldier (real case)

6.4 Toward a New Approach

When possible, all means usually summoned in the Toulouse region are for the search of bodies, whatever the environment, including parachutists of the Military Search. But in the cases that we had to deal with, all involving large forested areas, the field was not conducive to the deployment of the ground-penetrating radar. Some areas were difficult to access even for cadaver dogs and their handlers, but neither the environment nor the large surface area could be considered a problem, even if it was preferable to make relatively deep smell wells at regular intervals (at least every meter) over the entire research area. This was obviously not possible in the cases we are referring to in the Toulouse region. The twin dog handlers' methodology made it possible to cover the whole area. These interveners, such as the archaeo-thanatologist, whose work consists essentially of discriminating between bones that may have been discovered on the surface, directing the scraping and digging in the event of discovery, are therefore likely to highlight anomalies of relief to be tested. But the concentration on the dog's work and the permanent reading of his reactions do not really allow him to be interested in the different peculiarities of the field. That is why it was necessary to systematically involve specialists in reading the environment, knowing that we regularly accompany them in terms of training. We were therefore no longer alone to explore the entire research area on foot.

This approach, which is the beginning of the protocol we are trying to put in place, makes it possible, under difficult conditions, to manage lengthy research in large areas. The joint intervention of the various specialties, whose complementarity appeared to be obvious over the course of the cases, led to the testing of a common procedure to be efficient and rapid. The main idea is that the work of one entity should not be hindered by the exploration being carried out by another. This procedure is composed of ten phases (Éric et al. forthcoming), the originality of which lies in the creation of corridors, allowing the multidisciplinary intervention of the actors, and guarantees the quality of investigations by ensuring that they have actually covered the whole area. In the forest, with a relatively large number of operators on an area that it is not possible to observe from a single point of view, it would be easy to inadvertently forget parts of the land. The installation of a specific signaling system, according to traceability criteria, allows an immediate reading of the current operation.

The characteristics of mission preparation (phase 1), an essential step that is not limited to the material aspects and is all too often neglected (taking into account environmental data, prospecting by aerial imagery, etc.), as well as the coordination and designation of a control point (phase 2), designated under the acronym ICP for Incident Control Point, correspond to military requirements but are adapted to the forensic context of France. The designation of a control point partly responds to the recognition of the zone (phase 3), the first real act of field research. It is led by the scribe and group leader to see and determine where the research will begin and stop, depending on the exchanges with the investigator. However, it is desirable that representatives of all the intervening entities be present so as not to generate queries during the operation. This preliminary examination of the research sector also confirms the nature of the soils and vegetation, which are apprehended during the preparation phase (see above), and even confirms the use of various means of action besides defining priorities.

The delimited area is then marked (phase 4) (Fig. 6.8). The corridors are marked by yellow braid, denominated with a letter and numbers according to a system based on geolocation; the GPS coordinates of the ends of the lanes are systematically recorded and plotted graphically.

The size of the lanes may vary according to the configuration of the field and, above all, the length of the plot. The average width, generally equivalent to 5–6 m, is a good compromise between the methodology of dog handlers, who pay attention to the physiological requirements of their dogs (cf. *infra*), and the need to operate the scraping machine for checking anomalies. The action of the cadaver dogs is a delicate moment; they work first (phase 5) and alone for a while, until several corridors have been checked. It is indeed recommended not to disturb them by working close to them and, above all, not to come before them. Manual or mechanical checks would generate clods or piles of soil that could visually and olfactorily disturb the dogs. The size of the corridors also allows each dog handler to process it at once. Indeed, given the concentrated effort that this type of search requires from the dogs, each master carries out a rotation with his own dogs, who work only a few tens of minutes at most in one rotation. Their effort appears to be intense.

With at least one corridor left free of any intervention between them and the one where the dog and its master are working, the soldiers of the Military Search team

Fig. 6.8 Corridors (real case)

enter into action with their different metal detectors (phase 6). This knowledge on the detection of metallic objects present in the ground, with possible discrimination according to the nature of the metal, aims to find the victim according to the metallic elements that could be on or in his body.

In order to organize the research and facilitate the reading of the progress of work for all staff, a signage system based on a color code is systematically implemented. This signaling makes it easier for the investigator to resume work when the search is interrupted and the anomalies observed are located, as well as for the investigator to instantly read the current search operation, which is not insignificant. Indeed, this type of operation is often an opportunity for the latter to take time off to review different places of the case or to exchange with protagonists or interlocutors, especially when the surgery lasts a week. This signaling also indicates where the detected anomalies are located and whether excavations (Phase 7) are in progress.

The last three phases concern the topographic surveys (phase 8), the rehabilitation of the area (phase 9), and the report of the search operation (phase 10).

The Military Search team chose fiberglass stakes that do not disturb the action of the detectors because of their composition. They are arranged on the starting base and in the middle of the corridor so that they are visible to all. For marking of detections and excavations, there is a picket line at the beginning of the corridor and four pickets that mark the area of interest. Unmarked corridors downstream of the work area are considered to be made. The others, located upstream from the white stake ("corridor not made"), remain to be investigated. The selected colors are visible from afar, even with a relatively dense vegetation cover.

6.5 Conclusions

The search for buried bodies is a complex task, regardless of the environment in which it takes place. This is even truer when the place envisaged for the illegal burial of the victim is in the forest or in a place not really localized (research on large surface) and when the years have passed. The intervention of many specialists from different backgrounds is, without a doubt, the key to success. Unfortunately, this is rarely the case in much simpler cases, despite the fact that the IRCGN sends in specialists.

In the context of research on areas, often several hectares in the Toulouse region, a new approach has therefore been put in place, with the idea of imposing it on a national scale. The aim is to manage the simultaneous intervention of different specialties although no protocol currently exists. It will allow each person to apply his or her own methodology without compromising the effectiveness of the other. It is in this sense that the approach presented here has been thought out and adapted; it unquestionably makes it possible to work in the best conditions without creating harmful disruption for the continuation of investigations by other specialists. Above all, the marking of the corridors ensures that the entire surface is treated by each of the specialists, making it possible to follow the progress of research and the various interventions (residual human remains search dogs, metal detection, archaeological stripping) live, by means of a color code, even in dense environments and on very large surfaces.

Chapter 7
Actualistic Experimental Taphonomy of Inhumation Burial

Hayley L. Mickleburgh

Abstract Actualistic experimental taphonomy of inhumation burial represents an important research field with a wide range of applications in the forensic sciences, from improving postmortem interval estimation of buried remains to the detection of mass graves using remote sensing. Observational data on these processes and the effects of different variables could potentially be of great value in forensic contexts, e.g., to help determine whether there was postmortem human alteration of remains at a scene.

Keywords Actualistic experimental taphonomy · Funerary taphonomy · Archaeothanatology · Inhumation · Joint disarticulation sequence · Bone displacement · Body treatment

7.1 Introduction

Actualistic studies have a long history in archaeology, taphonomy, forensic anthropology, and paleontology (Bass 1997; Behrensmeyer and Kidwell 1985; Gifford-Gonzalez 1989; Lyman 1994; Pobiner and Braun 2005; Shirley et al. 2011). They are based on the assumption that contemporary knowledge can be applied to understand the past (uniformitarianism) and that there is a relationship between process and product (Rudwick 1971). Actualistic experimental taphonomy has become an important branch of forensic taphonomy, with a growing number of hypotheses being tested under controlled conditions. Important areas of experimental taphonomic research are the postmortem interval and modifying factors to human remains and evidence, for example, scavenger modification. The taphonomy of inhumation burials is an important line of investigation in both forensic and traditional (funerary) archaeology. For both fields, understanding the formation processes of human burials and distinguishing the effects of human manipulation of the body and its

H. L. Mickleburgh (✉)
Faculty of Archaeology, Leiden University, Leiden, The Netherlands
e-mail: h.l.mickleburgh@arch.leidenuniv.nl

P. M. Barone, W. J. M. Groen (eds.), *Multidisciplinary Approaches to Forensic Archaeology*, Soil Forensics, https://doi.org/10.1007/978-3-319-94397-8_7

surroundings from taphonomic processes are key to reconstructing the sequence of events surrounding the death and burial of an individual.

Experimental inhumation burials in forensic studies have focused both on surface placement of bodies and burial, although the majority of the studies has dealt with the former, in order to observe the stages of gross decomposition and to reflect common modes of disposal of remains by perpetrators. To date, experiments on the taphonomy of inhumation graves have made important contributions to knowledge of, i.a., the rate of decomposition in buried bodies (Marais-Werner et al. 2017; Rodriguez and Bass 1985; Wilson et al. 2007), biological alteration of the burial environment and soil (Tibbett and Carter 2008; Thomas et al. 2017), biodegradation of human tissues (e.g., Wilson 2017), degradation of clothing in buried environments (Stuart and Ueland 2017b), differential decomposition in mass graves (Troutman et al. 2014), and detection of mass graves using remote sensing (Kalacska et al. 2009). Body analogues for humans are used in many studies. Most often pigs (*Sus scrofa*) are used, as they are considered anatomically and physiologically similar enough to humans to serve as analogues, although other studies have used rabbit or rat analogues. Acquiring sizable samples of human bodies is a continuing challenge. The use of pigs as body analogues allows researchers to acquire larger samples and provides much greater control over variables such as body size, cause of death, and pathology, as well as speed of placement after death. In addition, pigs can function as human proxies, when the use of human body donation is not an option due to ethical concerns. However, recent research has indicated that nonhuman body analogues do not reflect the range of variation observed in humans. In a study comparing five humans, five pigs, and five rabbits, the researchers found that humans exhibited greater variability in both visual decomposition changes and insect activity than pigs or rabbits (Dautartas et al. 2015; Steadman et al. 2016). In another study comparing decomposition in 17 pigs and 22 humans, Connor et al. (2017) found that the trajectory of decomposition between the two groups diverged in both rate and gross presentation. They conclude that pigs may be useful in studying general trends but are not a substitute for human subjects. They further call for caution when applying data derived from pigs to human subjects in medicolegal investigation.

Data gathered from forensic case reports has also provided substantial information on the taphonomy of inhumation burials, and analysis of datasets large enough to provide statistically significant results offers important insights into the effects of different taphonomic variables, including (plastic) wrapping of the body, burial depth, and soil type (e.g., De Leeuwe and Groen 2017). In addition, such studies may arguably reflect real-world scenarios more accurately. On the other hand, experimental studies provide the important benefit of regulation of variables throughout decomposition and are designed to test specific hypotheses (which in turn can be developed based on forensic casework) (Sorg et al. 2012; Simmons 2017).

Recently, experimental taphonomy has been used to improve the analysis of inhumation burials from (non-forensic) archaeological contexts (Mickleburgh and Wescott 2018). Approaches to the taphonomy of the burial environment in traditional archaeology have been developed with the aim of distinguishing the multitude of taphonomic factors modifying human remains, and to understand past funerary practices and concepts of death and afterlife.

7.2 Funerary Taphonomy

In funerary archaeology, a subfield of traditional archaeology, the taphonomy of the human body within a grave is receiving increasing attention. Funerary taphonomy, recently defined as "the study of how taphonomic changes aid the interpretation of funerary practices" (Knüsel and Robb 2016: 655), uses the principles of taphonomy to reconstruct the sequence of events surrounding death and burial and to distinguish taphonomic alterations from human modification of human remains. The scope of this developing field is broad: in order to understand the social and cultural aspects of burial, it is important to be able to distinguish the effects of a multitude of taphonomic processes on assemblage patterning. These include, inter alia, bone modification (e.g., weathering, rodent gnawing, burning), mode of burial (e.g., single, primary inhumation, commingled deposition), decomposition, and body condition upon deposition (e.g., desiccation/mummification). The primary concern of funerary taphonomy is to identify patterning resulting from human funerary practices.

An important methodological approach used is *archaeothanatology*. This French- developed field methodology uses the spatial patterning of bones in a deposit, along with knowledge of anatomy and human decomposition, to draw inferences on the treatment of the body and its original position upon interment. The principle underlying this field is that the burial mode (e.g., grave shape and size, body position, presence of a body container, stage of decomposition of the body upon burial) will affect how the body decomposes and is preserved (Castex and Blaizot 2017; Duday 2009; Duday et al. 1990, 2014).

Archaeothanatology relies on principles of forensic taphonomy and an understanding of human anatomy, to interpret a human burial within its context. By analyzing individual bone position and using models of decomposition and the order of joint disarticulation, the original conditions of the deposit, the sequence of taphonomic events and human modifications are reconstructed. Skeletal disarticulation patterns can be diagnostic of environmental conditions (e.g., water flow), animal behavior (e.g., scavenging), and/or human action (e.g., ritual displacement of bones), aiding the reconstruction of the events that formed a burial feature. Final bone position is influenced by body position upon interment, open spaces and supporting structures in the grave, human/animal activity, soil conditions, and patterns of natural joint disarticulation. Archaeothanatology has made extensive use of taphonomic models of the natural sequence of joint disarticulation to distinguish natural from anthropogenic or bioturbation/scavenging processes. In archaeological cases, the application of these methods has allowed researchers to establish a timeline and the sequence of events surrounding the death and burial of individuals. In a study of Moche burial practices in pre-colonial Peru, taphonomic analysis of the disarticulation and spatial distribution of parts of skeletal remains demonstrated that bodies were (partially) mummified prior to burial, allowing disarticulation of brittle parts of the body during placement in a tomb. The analysis further indicated that there was a protracted time between death and final burial (Nelson 1998). Taphonomic analysis of medieval single graves in France has demonstrated the presence of perishable structural supports in the grave (wooden shelf, coffin) as well as revealed the shape and dimensions of the original grave pit. Analysis of post-medieval mass graves allowed the detailed reconstruction of the relative chronology of deposits within the feature, demonstrating that while the deposit was composed of several layers of individuals, the mass grave and all individuals within it were deposited in one event (Castex and Blaizot 2017). Archaeothanatological analysis furthermore helps to i.a. (1) determine if burials are primary or secondary; (2) distinguish between intentional and accidental burials, e.g., in paleoanthropology; (3) reveal whether the body was moved/transported before burial; (4) identify whether a burial has been disturbed post-deposition; and (5) infer the presence of containers and/or wrapping of the body upon interment (which have since decayed).

In the absence of actualistic taphonomic data on the factors influencing final bone position in the grave, archaeothanatology has relied on inferences resulting from repeated observation of patterns in the archaeological record. While archaeological burials can provide insights from a broad range of contexts and environments, archaeological contexts do not provide the possibility to control conditions and are the result (a "final snapshot") of a multitude of processes. As highlighted by Appleby (2016: 21), "there is a danger of circular argument: a burial is primary because of the persistence of labile articulations and as secondary if they are absent. At the same time, the articulations are identified as labile due to their frequent preservation in primary burials and absence in secondary ones. This means that lists of labile and persistent articulations are mostly based on inference about the way in which ligaments decompose from repeated experience." In fact, the comparative analysis of a large number of archaeological burials in supine position from different spatiotemporal contexts found little supporting evidence for the existence of a "common" or "natural" joint disarticulation sequence in humans (Peressinotto 2007), underscoring the need to establish how different variables affect disarticulation and spatial patterning of bones within the grave.

Actualistic experiments provide the opportunity to study the effects of specific conditions and taphonomic processes. Recent actualistic taphonomic experiments by the author aimed at collecting observational data on joint disarticulation and bone displacement throughout decomposition, and the effects of body position and condition on final patterning of skeletonized remains, explore the potential of actualistic experiments to improve and further develop archaeothanatology.

7.3 Actualistic Taphonomy Experiments for Archaeology

Decomposition is a continuous process of chemical, physical, and mechanical changes to the body. Decomposition processes such as bloating are known to cause movement of the body, especially the limbs. Once skeletonization and decomposition of connective joint tissues occur, individual bones can become displaced out of anatomical position. Five main factors are known to influence the final position of the skeletal remains: (1) body treatment and body position; (2) grave architecture and furniture; (3) stage of decomposition upon interment, decomposition processes, and the sequence of disarticulation of the skeleton; (4) environment, e.g., soil type; and (5) human and animal activities in and around the grave/body.

Recent experimental research by the author examines skeletal disarticulation and bone displacement during decomposition of five willed, donated human bodies. The pilot study (2015–2019) explores how different variables influence disarticulation and bone displacement, to improve interpretation of archaeological human burials, and aims to set parameters for future experiments with a larger sample. The overarching goal is to gain a better understanding of the formation processes of archaeological human burials through actualistic study under controlled conditions.

The remains were placed at the Forensic Anthropology Research Facility (FARF) managed by the Forensic Anthropology Center at Texas State (FACTS) University in Central Texas. FACTS receives whole body donations for scientific research under the Texas revised Uniform Anatomical Gift Act (National Conference of Commissioners on Uniform State Laws 2009). Body donations are exclusively acquired by Texas State University through the expressed and documented willing of the donors and/or their legal next of kin. Donors and/or their next of kin are aware that body donations are used for taphonomic studies. Demographic, health, and other information are provided through a questionnaire completed by the donor or legal next of kin. The program complies with all legal and ethical standards associated with the use of human remains for scientific research.

Photography and daily observations provided visual evidence to assess soft tissue decomposition, disarticulation of the skeleton, and bone movement. Photography also provided data for the production of 3D models using structure from motion. 3D models were created at weekly intervals to document and quantify movement of the body parts and skeletal elements over time.

7.3.1 Disarticulation Sequence

An important key to archaeothanatology is knowledge of the relative timing of disarticulation of the individual joints in the human skeleton. A model of the relative sequence of disarticulation is used to "distinguish the action of natural processes from those relating to the placement of the corpse as a part of funerary treatment" (Knüsel 2014: 30). The current understanding of the skeletal disarticulation sequence is based on the idea of labile (non-durable) and persistent (durable) joints (Duday 2009; Knüsel 2014). Labile joints are those that are relatively unstable and disarticulate relatively soon after death, and are frequently found out of anatomical position, unless there is supporting soil or burial architecture. These include joints held together primarily by soft tissue attachments (e.g., hyoid, mandible, patellae, scapulothoracic connection, costo-sternal joints) and small bones of the hands and feet (Duday 2009; Knüsel 2014; Roksandic 2002). Persistent joints, on the other hand, are major weight-bearing joints with ligaments that resist disarticulation and frequently maintain their anatomical position. This model of joint durability is based on (1) assumptions on durability during decomposition based on biomechanical function and strength in life and (2) observations of the anatomical relation of bones in archaeological burials. However, the validity of the model has been questioned based on contradictory findings in archaeological burials, suggesting there may not be a predictable disarticulation sequence based strictly on joint type (Peressinotto 2007).

While the potential effects of different body positions and body treatment (e.g., mummification) on the relative order of joint disarticulation have long been acknowledged (Duday et al. 1990; Maureille and Sellier 1996), the recent actualistic pilot study conducted by the author at FARF of five bodies in different positions suggests that the process of disarticulation is highly complex. The case study involved multiple instances of disarticulation followed by re-articulation, a phenomenon that could not be ascertained based on observation of archaeological burials, demonstrating that the current model may not capture the entire process (Mickleburgh and Wescott 2018). While the current disarticulation model is based on an assumed order of decay of the connective tissues of the joints leading to bone displacement, the actualistic study demonstrated that displacement of bones out of anatomical position is not necessarily dependent on complete decomposition of the connective tissues. Displacement was found to occur in joints still retaining (stretched) connective tissue. In addition, it remains unclear whether the strength or biomechanical function of individual joint articulations in life determines their durability during decomposition, which is one underlying assumption of the disarticulation model currently used.

7.3.2 Bone Displacement

The position of the bones can be critical to understanding the original body position, especially when perishable containers or burial architecture are no longer preserved. Gravity plays a major role in the disarticulation of joints and movement of individual bones depending on the position of the body. If the body is situated in a relatively stable position, little to no movement of the bones may occur. However, if the body is in an unstable position, the bones can move within open spaces due to the effects of gravity. This process is known as *necrodynamics* (Wilhelmson and Dell'Unto 2015). Open space can be present upon deposition of the body, e.g., in a coffin (primary open space), or may be created by decomposition of the soft tissues and burial architecture (secondary open space). Previous studies of archaeological burials have indicated that different body positions may affect the order of joint disarticulation (Gerdau- Radonic 2012; Peressinotto 2007; Ortiz et al. 2013; Richter et al. 2010; Rottier 2016). Understanding bone displacement, its relation to the sequence of disarticulation, and the presence or creation of open spaces can be critical to reconstructing the relative chronology of processes within a burial feature, as demonstrated by the analysis of post-medieval mass burials discussed above (Castex and Blaizot 2017).

To examine the effects of primary and secondary open space on disarticulation and bone displacement, two donated bodies were placed in small oval-shaped pits at FARF in upright, seated position. This body position, which is known from various spatiotemporal contexts in archaeology, was expected to lead to greater forces on upper parts of the body and a greater distance of displacement of bones, due to the effects of gravity and the superposition of body parts (Gerdau- Radonic 2012; Ortiz et al. 2013; Richter et al. 2010; Rottier 2016). Of the two donated bodies, one was immediately buried with soil upon placement, while the other was placed in a pit that remained open throughout the duration of the experiment. The experiment was continued until complete skeletonization. 3D models were developed weekly for the open-pit experiment, and upon placement and final excavation for the burial experiment. The sequential models can be overlaid to allow quantification of bone displacement, i.e., the direction and volume of movement, over time. The burial experiment was expected to produce less movement of the bones due to supporting soil. Furthermore, based on archaeothanatological research, the burial experiment was expected to show displacement of bones only within the soft tissue body volume, while the open-pit experiment could produce displacement outside of the soft tissue body volume.

The experiment revealed that the volume and pattern of bone displacement in the buried donation was similar to that of the open-pit donation. Despite the duration of the burial experiment (793 days), secondary open space was very well preserved, leaving a precise impression of the soft tissue volume of the body in the soil. Upon skeletonization, bones were able to move within this secondary space, and the amount of vertical displacement of bones was similar to that in the open-pit experiment. This highlights inferences based on archaeological burials that the formation of secondary open spaces is a major factor in the final distribution of bones in a deposit. The formation and maintenance of secondary open space is dependent on soil type and the weight of overlying soil, with sandy soils known to rapidly or immediately fill newly created voids (hourglass effect), and soils with a clayey texture preserving voids for a longer duration (Duday 2009). In the burial experiment, fine plant roots were also found to be a major factor in the preservation of voids in the burial pit. In addition, displacement outside of the soft tissue body volume was found to be minimal in the open-pit experiment.

These preliminary results indicate that further investigation of the role of soil type in the creation of secondary open spaces, as well as the duration of time in delayed infilling, is necessary. Overall, the results thus far support earlier findings that the upright seated body position leads to greater bone displacement, than in supine burials. Displacement outside of the soft tissue body volume was minimal in the open-pit experiment, and without repetition of the research with a larger sample, it is impossible to assess the efficacy of this observation as a marker to distinguish between the presence of primary and secondary open space throughout decomposition.

7.3.3 Body Condition Upon Interment

The condition of the body or stage of decomposition at burial can affect the final disposition of the bones. As Knüsel (2014: 46) notes, "delay in burial may occur for a number of reasons: corpse preparation, as in mummification or exposure/storage leading to desiccation; death at a distance; political manoeuvring to ensure succession; or disrupted or lacking ceremonies, planned or simply omitted (as with the dead from battle, which were often left exposed on the field of battle for varying periods of time without burial)." The potential volume of secondary open space in the burial pit created by decay of the soft tissues can differ significantly between bodies interred in different stages of decomposition. Fresh bodies, yet to go through bloating, or bloated bodies, will present larger soft tissue volume than desiccated/mummified bodies. As discussed above, the formation of secondary open space, or voids, allowing bone displacement is dependent on the decay of soft tissues or burial architecture and delayed infilling by the surrounding soil matrix. As such, desiccated body interments lacking primary open space would be expected to demonstrate less bone movement upon skeletonization than fresh or bloated body interments, due to the difference in soft tissue volume. In addition, desiccation is thought to affect the order of joint disarticulation. Cases of "paradoxical articulation" have been described in archaeological burials, in which skeletal remains show preserved labile articulations, but persistent joints have disarticulated. Paradoxical articulation has been suggested to be an indication of desiccation of the body, which limits bone movement and in combination with human manipulation of the remains causes deviation from the "normal" sequence of joint disarticulation (Maureille and Sellier 1996).

To examine the effects of desiccation of the soft tissues on the formation of secondary open space and bone displacement, as well as possible "paradoxical articulation," one donated body in fresh decomposition stage and one naturally mummified body were placed in small oval-shaped pits in supine flexed position. The fresh body donation was excavated after 834 days, at which time the body was found to be completely skeletonized. Secondary open space was well preserved, leaving a clear impression of the soft tissue volume of the body in the soil. Overall, the pattern of bone displacement differed from the upright seated experiment described above, as the supine body position allowed less vertical bone displacement. Displacement was most prominent in the lower limbs and thorax, and bones remained within the original soft tissue body volume. The mummified body experiment is ongoing at the time of writing. The state of articulation and bone displacement in this donation will be assessed upon excavation.

7.4 Application in Forensic Contexts

Forensic archaeology and archaeothanatology have two important aims in common: distinguishing taphonomic alterations from human modification of the remains and their context and reconstructing the sequence of events surrounding death and thereafter. This has led to calls within the field of archaeothanatology to apply its methods in forensic contexts (Castex and Blaizot 2017; Duday and Guillon 2006). While verifiable data collected under controlled conditions is highly valuable to the development of archaeothanatology and funerary taphonomy (as discussed above), it can be argued to be essential in order for this methodological approach to be applied in forensic cases. In a legal setting, it is important that forensic archaeologists can support their interpretations of the data with appropriate (robust empiric) evidence. Actualistic experimental data provide a framework of clear data collection protocols and controlled conditions to establish the role of specific variables. However, despite recent research collecting experimental data to support and improve archaeothanatological methods, the potential for its application in forensic contexts is still impeded for a number of reasons. Foremost among these is the lack of large sample sizes in experimental studies. In a forensic context, where information derived from experiments may need to be defensible in court, statistically significant findings are important. For a solid scientific foundation of the experiment design, repetition and replication of experiments, and the use of control groups, are essential. Preferably, the number of controls in any given experiment would be equal to the number in the test scenario (Simmons 2017). Repetition of experiments reduces the role of chance variation in the results, making statistical inferences more reliable. Furthermore, since human decomposition is known to be highly variable depending on environment, replication of the experiments in different locations, reflecting different environmental conditions, is crucial. Human decomposition research facilities throughout the United States, the Netherlands, and Australia provide the opportunity to replicate experiments in different environments and compare results. However, acquiring substantial sample sizes of donated human bodies remains a challenge. The use of body analogues in human skeletal disarticulation studies is not a viable option, since observations of nonhuman disarticulation sequences indicate that they differ per species (Hill 1980).

7.5 Discussion/Conclusion

Actualistic experimental taphonomy of inhumation burial represents an important research field with a wide range of applications in the forensic sciences, from improving postmortem interval estimation of buried remains to the detection of mass graves using remote sensing. For archaeothanatology and the study of past funerary practices in general, actualistic experiments on human joint disarticulation, bone displacement patterning, and the role of a multitude of variables including body condition, body position, and primary and secondary open spaces represent an important opportunity to observe the entire process under controlled conditions, with the ultimate aim of establishing markers to distinguish between (intentional) human actions and natural processes during excavation. Observational data on these processes and the effects of different variables could potentially be of great value in forensic contexts, e.g., to help determine whether there was postmortem human alteration of remains at a scene. However, for both traditional and forensic archaeology, repetition and replication of experiments (in different environments) and the use of control groups are crucial, to allow reliable statistical inferences.

Chapter 8
The Use of Three-Dimensional Scanning and Surface Capture Methods in Recording Forensic Taphonomic Traces: Issues of Technology, Visualisation, and Validation

Patrick S. Randolph-Quinney, Stephen D. Haines, and Ashley Kruger

Abstract Three-dimensional (3D) space capture is now routinely applied in forensic practice. This has often taken the form of using pseudo-3D visualisations such as 360° photography (return to scene) or digital photogrammetry or true 3D space capture using laser scanning (to derive surfaces), or total station survey methods (to derive Cartesian coordinates). Often these are used to record topography and spatial distributions at crime scenes and may be used to provide a spatial archive of evidence found at a scene or as an aid in visualisation for courtroom purposes. However, there is a growing interest in the use of 3D data capture methods for recording and analysing taphonomic evidence, both for purposes of recording and data sharing, but also to facilitate formal taphonomic analyses which are often qualitative with regard to taphonomic trace criteria. However, as the application of 3D data in taphonomy is a relatively new phenomenon, there remains little consensus on what equipment and imaging modalities are either appropriate or indeed best, to use, and whether digital models of taphonomic traces are analytically valid or verifiable. This paper sets out to highlight and evaluate a number of technological approaches,

P. S. Randolph-Quinney (✉)
School of Forensic and Applied Sciences, University of Central Lancashire, Preston, Lancashire, UK

Evolutionary Studies Institute, University of the Witwatersrand, Johannesburg, South Africa

School of Anatomical Sciences, Faculty of Health Sciences, University of the Witwatersrand Medical School, Johannesburg, South Africa
e-mail: prandolph-quinney@uclan.ac.uk

S. D. Haines
School of Forensic and Applied Sciences, University of Central Lancashire, Preston, Lancashire, UK

A. Kruger
Evolutionary Studies Institute, University of the Witwatersrand, Johannesburg, South Africa

P. M. Barone, W. J. M. Groen (eds.), *Multidisciplinary Approaches to Forensic Archaeology*, Soil Forensics, https://doi.org/10.1007/978-3-319-94397-8_8

visualisation methods, post-capture processing methods, and analytical criteria for effective 3D data acquisition of taphonomic traces. We provide an overview of current trends and possible future directions in the application of 3D capture and imaging methods for taphonomic research and practice.

Keywords Forensic taphonomy · Palaeotaphonomy · Laser scanning · Structured light scanning · Photogrammetry · Computed tomography

8.1 Introduction

The term taphonomy was first introduced by the Russian geologist Efremov (1940b: 85) to encompass studies in post-mortem and post-depositional processes marking the 'transition of animal remains from the biosphere into the lithosphere', the word being derived from the Greek *taphos*, meaning 'burial', and *nomos*, meaning 'law'. Since then, the term has been variously applied to a wide range of research in palaeontology, palaeoanthropology, and archaeology – the latter often associated with site formational processes (Allison and Briggs 1991; Andrews 1995; Bartosiewicz 2008; Beherensmeyer et al. 2000; Behrensmeyer 1985; Brain 1981; Duhig and Martinsen 2007; Gifford 1981; Haglund and Sorg 1997; Hill 1984; Komer and Buikstra 2008; Lyman 1994; Lyman 2010; Olson 1980).

It is only in the last three decades that taphonomy has been adopted by the forensic community under the umbrella term *forensic taphonomy*, where it is considered a branch of anthropology or archaeology synonymous with all aspects of the perimortem, post-mortem, and burial processes (Haglund and Sorg 1997; Hochrein 2002; Nordby 2002; Simmons 2002; Sorg and Haglund 2002; Tibbett and Carter 2008; Tibbett and Carter 2009). The co-option of taphonomy into forensic science was described by Dirkmaat and colleagues (2008: 34) as 'the most significant development [to] alter the field of forensic anthropology' in the last 20 years (Dirkmaat et al. 2008). Forensic taphonomy is unique in that analyses have moved away from longitudinal studies of complex time-averaged assemblages accumulated over millennia to shorter post-mortem time frames spanning days to years, with the acknowledgment of humans as taphonomic agents and the emergence of the human cadaver as a key unit of analysis (Bristow et al. 2011).

There is a developing literature in the field, with several seminal textbooks available (Pokines and Symes 2014, (Haglund and Sorg 1997), and a diverse journal literature dealing with forensic aspects of taphonomic factors (see Bristow et al. 2011 for an overview). Over the last two decades, this literature has spanned a number of related scientific areas of taphonomic application, including the fields of forensic anthropology, archaeology, entomology, genetics, medicine, chemistry, biology, and engineering; there are today many differing scientific subdisciplines which together contribute to science under the umbrella of forensic taphonomy.

Epistemologically, the discipline has developed using a broadly uniformitarian approach (*sensu* Lyell) utilising the study of modern analogues as a means of interpreting observed past phenomena. Forensic taphonomic studies very often fall under the category of neo-taphonomy or actualistic taphonomy, which concentrates on the modern environment and applies its results to the past by analogy. This contrasts with 'classical' palaeo-taphonomy (*sensu* Efremov), which examines the context and content of depositional sites in great detail using temporo-spatial patterning, skeletal part representation, and the pattern of skeletal damage as a means of interpreting formation processes. The neo-taphonomic approach can be seen as primarily hypothetico-deductive in nature, whilst palaeo-taphonomy can be inductive, deductive, or a combination of both (Bristow et al. 2011). With these distinctions in mind, forensic taphonomic studies primarily fall into three broad categories:

1. Post-mortem interval estimation – this includes applications of forensic entomology, diagenesis, decompositional morphology, volatile organic chemistry (VOC), osseous changes associated with subaerial weathering, and studies of the microbiome.
2. Body disposition – this includes issues of biotic and abiotic burial processes (including aspects of weathering, sub-surface erosional effects, fluvial transport, and predation and scavenging) and interactions with the burial environment and its ecology.
3. Human intervention – anthropogenic impact on the human body, including the effects of trauma (blunt, sharp, hacking, and ballistic), dismemberment, anthropogenic dispersal, and thermal alteration and burning.

As such, there are many causative processes which can produce taphonomic changes on bone. Features of alteration to bone such as colour changes, surface modifications, or weathering and scavenging effects can be highly informative as to the environment and context of deposition during the peri- or post-mortem periods (Pokines and Symes 2014). From this basis, being able to accurately record, operationalise, and analyse taphonomic factors can be considered incredibly important in the forensic setting and may present significant evidential potential in the investigation of criminal acts or to corroborate or undermine witness testimony. It is thus imperative that such traces are recorded objectively. Given the breadth of research (and forensic case relevance) that taphonomy currently expresses, it is unsurprising that 3D digital recording and analytical methods have been applied to different aspects of taphonomic study.

8.2 3D Applications in Taphonomy: From Site-Space to Surface Modifications

This is a fast-developing area of both scientific and technological innovation, though in the main applications currently seem fall into two broad categories: (1) spatial visualisation and site-space analysis from the local deposition site up to landscape level and (2) visualisation and quantitative analysis of taphonomic alterations and modifications to small-scale portable objects. In both categories there are generally two covarying end products of any 3D application – visualisation, which provides a structured 3D model which may be scaled, rotated, and interrogated from multiple viewpoints in 3D space, and analysis, which allows aspects of either internal or external structure or morphology, or both, to be quantified and subject to appropriate statistical modelling.

3D recording of forensic archaeological deposits has been undertaken since the 1990s as part of standard archaeological survey of commingled or mass graves, from which the spatial relationships between individuals can be recorded (Adams and Byrd 2008; Herrmann and Devlin 2008; Tuller et al. 2008). Such approaches, using electronic survey systems such as a laser-based total station or differential global positioning system (GPS) to capture low-density Cartesian coordinates (LDCC) of objects as points in space, are then used as part of reconstructing and interpreting site stratigraphy. Whilst the use of laser or GPS methods can produce highly precise spatial coordinates, the inherent limitation of the LDCC approach is that it fails to record, visualise, and resolve surfaces (essentially the 'bits in between') which can be informative with regard to spatial relationships, morphology, and spatial complexity.

As such, there has been a shift towards surface (rather than point) capture in many areas of field practice. In the main these use non-contact methods of surface recovery (utilising laser or structured light scanning or photogrammetric methods – see Sect. 8.3 below) to generate either a point cloud or photogrammetric model of surface morphology. The technology for generating visually arresting and user-friendly accessible models based on such approaches (particularly photogrammetry) has been widely applied in recent years to archaeological and heritage sites. The end product of the heritage-driven approach is almost exclusively the generation and dissemination of an interactive model with which users (often non-specialists and the general public) can interact. Such models are generally a means to an end and require no deeper quantification or interrogation beyond their inherent aesthetic worth. Analytical approaches to such data have tended to fall under the umbrella of geomatics, within which 3D recording methods are considered digital, objective, rapid, and highly cost-effective compared to traditional nondigital approaches (Abel et al. 2011; Allard et al. 2005; Bello et al. 2009, 2013; Böhler and Marbs 2004; Breuckmann et al. 2009; Doneus et al. 2011; Galeazzi 2016).

However, recent approaches in the palaeosciences to the visualisation of prehistoric sites have been developed to allow quantitative recording of 3D systems, with express application to taphonomic analysis. At the Middle Pleistocene cave site of Rising Star in South Africa, excavators have applied multimodal recording systems (combining photogrammetry from drone UAV-mounted cameras, high-density point cloud laser scanning, and structured light scanning) to generate relational 3D models of cave and landscape morphology which have been used to interpret factors such as site formational processes and hominin mortuary behaviours (Dirks et al. 2015; Kruger et al. 2016; Randolph-Quinney 2015).

From this and similar projects has developed the subdiscipline of spatial taphonomy (see Domínguez-Rodrigo et al. 2017). Such approaches use powerful geostatistical tools based on 3D morphological data that reinforce inferences and hypothesis testing. The field has recently emerged in response to the need for a specific taphonomic perspective in spatial analysis. The application of quantitative models and spatial statistics overcomes the traditional, intuitive, 'eye-balling' methods of interpreting spatial patterns which enable researchers to draw statistically unsupported inferences about the taphonomic processes responsible for an observed pattern. Whereas some of the analytical tools of spatial taphonomic studies have long been tested and applied (i.e. fabric analysis), other methods (such as geostatistics or point pattern analysis) have recently demonstrated great versatility and validity (Domínguez-Rodrigo et al. 2014, 2017). These techniques are readily applicable to forensic studies, and are likely to develop significant importance in the future, particularly in relation to understand mortuary practices, decomposition, and transport in taphonomic contexts.

At the other end of the spatial scale, 3D recording of small objects (e.g. manuports, tools, or bones) is becoming increasingly common both in terms of visualisation (as the interactive end product) and quantitative recording (Errickson 2017). This has ranged from using structured light or laser scanning to facilitate the recording of overall normal morphology or pathological conditions on bone (Errickson 2017; Wilson et al. 2017); the use of 3D laser scanners to document ephemeral evidence at crime scenes, post-mortem examinations, and trauma (Dittmar 2017; Ulguim 2017, and see Fig. 8.1); visualisation and quantification of abrasion on water-submerged bone (Griffith and Thompson 2017); or the spatial position of skeletons within mass graves which bridges the scales between site-based and object-based recording (Baier and Rando 2016).

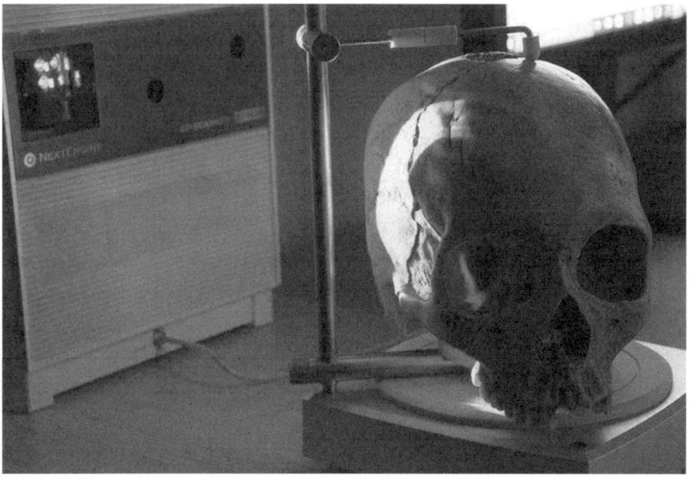

Fig. 8.1 NextEngine tabletop 3D laser scanner in operation, showing skull with cranial trauma being scanned on the turntable

8.3 Modalities of 3D Capture

There are numerous methods of recording and visualising objects in three dimensions (termed shape acquisition), including optical, tomographic, and contact-based methods. With regard to precise capture of internal and external structures, methods based on the focussed transmission of X-radiation are generally considered to represent the 'gold standard' – in particular, volumetric data are based on image slices from computed tomographic (CT) scanners. CT produces a three-dimensional image volume from a series of two-dimensional X-ray images. The individual image slices from a CT volume are used to produce a virtual three-dimensional object, through a process known as 3D reconstruction.

The most important aspect of this process is the application of rendering to the image volume. Rendering involves discrete stages by which computer algorithms process the geometry of an image stack, through transformation, perspective creation, clipping, and lighting, and in doing so create an image based on rastering, shading, texturing, and anti-aliasing. From this rendering process, two modes of images may be produced – a surface-rendered model or a volume-rendered model. A surface-rendered model (or isosurface) represents points that have equal values of grey density which are then extracted and extrapolated as a series of polygons into a proper geometric surface, for instance, the external surface of a bone. Volume rendering, on the other hand, treats grey values as being partially opaque, allowing the observer to see into or through a solid structure to a greater or lesser extent. In practice isosurface rendering is the preferred option for subsequent interrogation of surfaces, and isosurface files can be exported as relatively small digital models in common formats such as PLY or STL for subsequent manipulation and landmarking using proprietary or open-source software (now built into standard operating systems such as Microsoft Windows 10 or available from open-source developers).

Other methods of surface (rather than structural) scanning are available, and these are primarily either based on optical methods (such as photogrammetry or using structured light) or laser reflectance methods. These may record surface geometry (shape), texture (colour or detail), or both. In most cases the process of model generation follows a typical workflow (Fig. 8.2) which starts with the production of a point cloud (x, y, z coordinates) of varying densities. Once collected, point clouds can be outputted as individual data files with x, y, and z coordinates, allowing an intermittent 3D representation a scanned object's surface to be created (often visualised as a ghostlike representation). Following cloud generation, the point data is typically converted into polygon and triangular mesh formats, onto which a rendered surface can be applied, whereby recorded vertices (points) in a cloud are connected together into a continuous 3D surface approximation. The final model may be decimated or down-sampled, smoothed, polished, and simplified in order to deaccentuate or remove extraneous features or to make the model small enough (in file size) to be used in a variety of software packages.

Of the three most common shape capture methods, photogrammetry starts from the acquisition of multiple images of a subject taken from different angles, typically

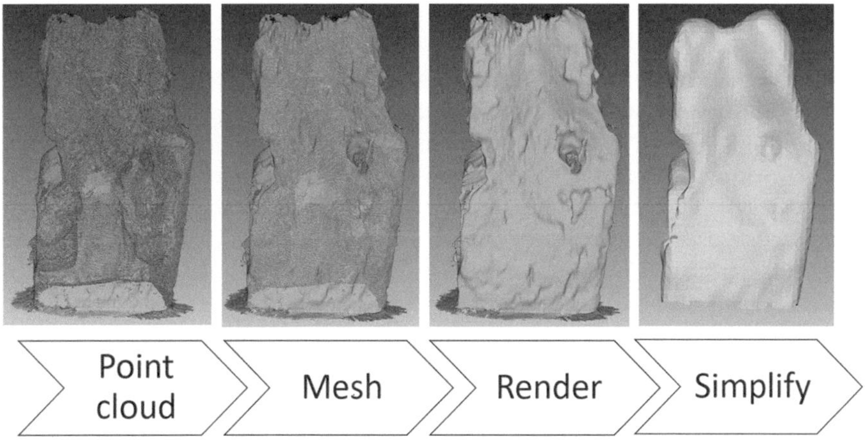

Fig. 8.2 The process of 3D model construction from point cloud data

using a DSLR (digital single-lens reflex) camera. The geometry of the subject is then calculated separately via several mathematical algorithms – the most common are SfM (structure from motion) and MVS (multi-view stereo) which require good coverage of the object to be scanned. There generally must be overlap between successive images in order to detect and define feature points (congruent points). These unique points, represented in two or more images, are used to reconstruct the scene geometry. Software then stitches together the combination of images to create a full 3D model and optionally a point cloud if required. This point cloud can also be transformed into a closed mesh with surface render for a more complete visualisation,

Structured light scanning (SLS), on the other hand, is based on the principle of projecting a narrow band of light onto a three-dimensionally shaped surface. This produces a line of light that appears distorted from other perspectives than that of the projector and which can then be used for geometric reconstruction of the surface shape. SLS typically consists of a camera with a mounted projector and uses patterned light through a form of geometrical calibration to generate its image. Such a projection method uses incoherent light and basically works like a video projector. Patterns are usually generated by passing light through a digital spatial light modulator, typically based on one of the three currently most widespread digital projection technologies, transmissive liquid crystal, reflective liquid crystal on silicon (LCOS), or digital light processing (DLP; moving micro-mirror) modulators, which have various comparative advantages and disadvantages. There are depth cues contained in the observed stripe patterns, and the displacement of any single stripe can directly be converted into 3D coordinates. 3D digitisation is completed by analysing the bend of the projected lines when visualised on the surface of the object. As such, the method is an active light documentation technique which digitises objects non-invasively, preserving the surface in its entirety.

Laser reflectance 3D surface scanners provide high-resolution representations of an object's surface. Such scanners generate a point cloud of subject geometry using reflected energy in the form of laser light. A laser beam is projected onto the surface of the object and reflected to a sensor element (usually a CCD camera). The relative positions of the laser source and the sensor array are known (through magnetic inferometric registration, applied targets, a fixed position turntable, or the use of a 5 or 6 degree of freedom armature), and the distance from the laser to the object can be triangulated depending on which sensor cells in the CCD are activated by the reflected laser light, allowing the camera to record cross-sectional depth profiles. In practice many handheld scanners acquire three-dimensional surface images when a laser scanning wand is swept over an object, in a manner similar to spray painting, or with the object mounted on a rotating turntable, with the object's image immediately appearing on a computer screen. Because such scanners provide real-time visual feedback, monitoring and controlling the scan process are relatively straightforward. When scanning complex shapes, more than one set of sweeps may be necessary to adequately record all the surface morphology, and thus individual sweeps of point clouds must be aligned or registered into a coherent 3D surface model using software dedicated to the device. Some laser scanners allow for the overlaying of photogrammetric texture data directly onto the generated polygon or mesh, providing surface colour and object detail in addition to geometric surface properties.

8.4 3D Taphonomic Applications: Issues in Visualisation and Model Validation

There remain few studies from which to assess best practice and best technology in using 3D scanning for taphonomic purposes – what studies have been undertaken being tied to specific technological devices (English Heritage 2011; Errickson et al. 2014). As such, it can be extremely difficult for practitioners to find objective advice when considering which modality to use, which equipment to purchase, and what post-scan software to adopt. In considering such questions for taphonomic recording, there are several options readily available of varying device type and cost, ranging from the relatively inexpensive method of photogrammetry using cheap (<$800) DSLRs, intermediate-cost desktop laser scanners (c. $3000), high-cost armature laser scanners (>$30,000), and handheld structured light options (>$15,000). However, regardless of the precise modality and specification of scanner adopted, there are a series of issues which should be born in mind before embarking on surface scanning for taphonomic purposes.

The choice of scanner and scanning modality being used will very much be influenced by the nature of the object required to be scanned. There are several considerations to bear in mind when choosing the optimal scanning method and technology, which include issues of relative cost, technical specification and performance, and the physical properties of the scan subject itself. By way of illustration we use

Table 8.1 Surface capture modalities and equipment used in following image examples

Modality	Equipment
Photogrammetry	Using a Nikon D5100 DSLR fitted with an 18–55 mm lens. Three lamps were placed at different angles around the subject to eliminate any potential shadows. Black velvet was used to create a homogenous background as an aid in masking the models. The camera position was fixed, and the object rotated on a turntable. Images were processed using Agisoft PhotoScan to build a sparse cloud, from which a dense point cloud is extrapolated, and from this a mesh (geometry) and texture model was produced
Laser scanning	Using a NextEngine Ultra HD tabletop scanner (http://www.nextengine.com). This device uses multistrip red light lasers to record surface geometry, with a reported dimensional accuracy between ±100 and 300 μ in macro and wide modes, respectively. Unlike many laser scanners, the NextEngine can capture texture utilising twin 5.0 megapixel cameras
Structured light scanning	Using a handheld Artec Eva SLS device (https://www.artec3d.com/portable-3d-scanners/artec-eva). This scanner has a reported 3D resolution of up to 0.5 mm with a point coordinate accuracy of 0.1 mm and distance precision of 0.03% over 100 cm

real-world examples and contrast three modalities of surface capture using affordable portable technology (Table 8.1) and briefly examine how different portable 3D capture modalities preform at recording taphonomic traces – considering price, user friendliness of the hardware and software, and geometric/textural model quality in the capture of common types of taphonomic modification on bone; this may include burning/charring, carnivore modification, subaerial weathering, ballistic trauma, or cut marks.

8.4.1 Preprocessing Issues and Technology of Surface Capture

There is considerable room to introduce error during the physical setting up of the scanning environment in relation to the imaged sample – as such these issues fall into intrinsic (sample-based) and extrinsic (scanner-based) considerations.

Intrinsic issues arise from the nature of the subject itself. As different 3D capture modalities have varied limitations, these must be taken into consideration to allow for optimum surface capture and the production of an accurate model. Intrinsic properties can include:

- Size of the subject
- Size and surface density of taphonomic modifications (homogenous or heterogenous)
- Complexity of the modifications (simple linear striae, crazing or patination, complex fracture patterns, etc.)
- Overall colour of the subject (dark, neutral, light)
- Presence of patches of surface colour

- Complexity and morphology of edges (rounded, thick or thin edges and fracture planes)
- Level of surface reflectance (matt, smooth sheen, or polished/reflective)

Extrinsic factors concern the capture modality itself and/or the inherent specifications and limitations of the scanning technology itself. Factors can include:

- Wavelength/colour of emitted light
- 3D resolution
- 3D point accuracy
- Texture resolution
- Linear and angular fields of view
- Working distance
- 3D depth resolution at distance
- Data acquisition speed
- The number of overlapping photographs, scan sweeps, or scan divisions (overlapping panels of scans which are manually or automatically registered and aligned to make a whole model from composite parts)

Recognising and dealing with these issues is often common sense, tempered with consideration of the scanners technical ability in relation to the environment it is being used in. Factors such as the nature of ambient lighting, leaving room for moving parts or turntables to operate normally, optimising the distance between capture device and sample, and the likelihood that the sample is within the technical capability of the scanner (based on sample colour, size, morphology, etc.) are all significant.

For instance, the zone of focus of the scanner needs to be considered when determining the position of scanning and alignment of the specimen if using texture capture in additon to attainment of geometry – if the sample is out of focus (blurred) within the scanners optical system, then surface capture will result in an out of focus model in post-processing which cannot be rectified. Ensuring there is enough room for the sample to be in focus across the entire scan process is paramount as this error will permeate throughout the production of the model. In situations where the scan device is fixed and the subject rotates (as in the use of the NextEngine), the subject ideally needs to be as close the centre of rotation of the turntable as possible – extremes of variation in the distance from optics to subject may cause parts of the subject (such as the ends of a long bone if mounted horizontally) to move in and out of the plane of scan focus. The physical stability of the sample also needs to be considered, both for the conservation of the sample (damage may result if it falls off a turntable for instance) and for stable scan capture. If the sample is unstable during scanning, it may move or wobble during recording, leading to capture becoming blurred or incorrectly aligned during post-processing.

It is worth noting that many common complications in scan quality are likely to be due to human error rather than failure of the capture device. Instrumental errors can cause the large inaccuracies when scanning objects, and correct calibration of the scanner is known to reduce or eliminate most of these errors, and it is beholden on the user to ensure correct calibration is undertaken.

8.4.2 Scan-Based Effects and Issues of Post-processing

All surface capture regardless of the specific technology introduce some level of error and imprecision into the production of finished models. Some of these are due to intrinsic and extrinsic effects noted above, whilst others are caused or exacerbated by the computational processes whereby point clouds are generated, meshes are produced, and where sets or sweep of scans are aligned and registered into a single shape volume. There is also a degree of error caused using post-processing affects such as smoothing, hole filling, editing, simplifying, and down-sampling of complex mesh data that influences the resultant quality of the model.

Probably the greatest problem in the production of surface models stems from the interaction between emitted energy (in the case of structured light or laser scanning) and either the colour, reflectance level, or degree of porosity and concavity (the presence of holes or deeply concave areas) expressed by the subject. These create problems in several ways, most commonly in the production of holes or voids in the scan:

- Reflectivity. A shiny reflective surface causes emitted energy to reflect or bounce, which introduces considerable interference on the surface of the point cloud. This can present as peaks or spikes, resulting in a 'noisy' model when meshed and rendered (Fig. 8.3a).
- Colour. If the colour of the object is too dark, then the laser or structured light beam is heavily absorbed, as is commonly seen in cases of soil staining to bone

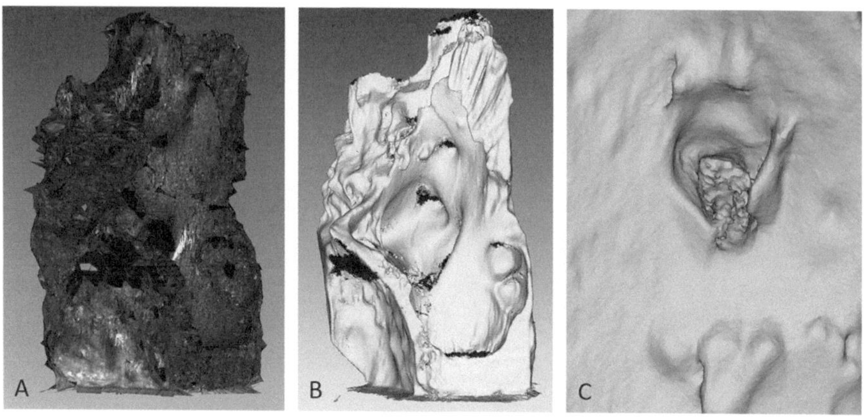

Fig. 8.3 Examples of surface degradation effects using a burnt bone example. (**a**) Spikes and artefact projections. (**b**) Hole defects and missing data. (**c**) Fill defects and scan interference. Scans undertaken with Artec Eva

or in charred/burnt remains. Very dark objects will present a series of holes, which will require filling (an approximation method) and may result in considerable data loss (Fig. 8.3b).

- Concavity. Holes, foramina, and deep concavities prevent some or all of the emitted energy from being reflected. Rather than being completely absorbed, this causes energy (particularly laser) to 'bounce around' inside the concavity, causing localised interference and creating false morphology (Fig. 8.3c).
- Opacity or translucency. This may cause reflectance issues below the immediate surface of the object, generating noise in the process. In highly translucent objects such as fresh greasy bone, this may also create a false (sub)surface far removed from the actual extant surface morphology. This problem is particularly acute in high-resolution photogrammetry and may end up producing an unusable false model as the consequence of 'seeing' below the true surface of the object.

Problems also arise in the scanning of thin/sharp edges, which may create reflectance issues at fracture margins. This is further exacerbated by problems of alignment in thin edges, which create artefacts along the margin (Fig. 8.4). This is problematical in cases of recording blunt force or ballistic trauma, where fracture morphology and topography (particularly the nature of inward-facing versus outward-facing edge bevelling) are vitally important in analysis and interpretation.

Finally, the post-processing workflow itself can introduce significant artefacts or lead to information loss. The requirements of taphonomic recording require that processing artefacts are kept to a minimum (limiting the creation of false morphology) and that information loss is minimised (limiting holes or missing data). In all cases the need to optimise surface capture should be recognised at the outset. This can be facilitated by creating significant overlap between scans of an object, with

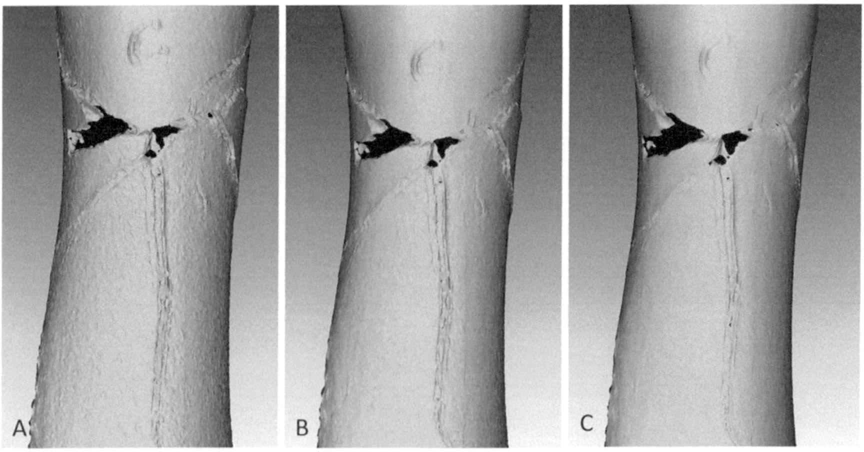

Fig. 8.4 Examples of surface degradation effects on NextEngine-scanned sample at standard resolution. (**a**) Raw scan. (**b**) Scan following mesh simplification. (**c**) Mesh smoothing following simplification

multiple angles (lines of sight) covered across the object shape space. This is achieved by taking multiple scans or scan sweeps, with overlap between scan passes allowing for good alignment.

However, this process results in much duplication of data with the result that models are decimated (down-sampled) to reduce the size of the final point cloud before creating a surface mesh. This, in and of itself, is not an issue; providing sufficient registration points exists between each scan and sweep. Most commercially available scanners feature an auto-align function in their operating software, though it should be noted that auto-alignment (or automatic registration) is a computational best-fit process and may not produce precise registration of scans, particularly in cases where topographical features are limited on the surface. In such a case, the use of manually affixed registration markers on the surface of the subject, coupled with manual registration of scans before fusing them into a whole, may be preferable. Paradoxically, using the highest available resolution in attempting a scan is not always the best option. Figure 8.5 illustrates the need for user-assessed optimality

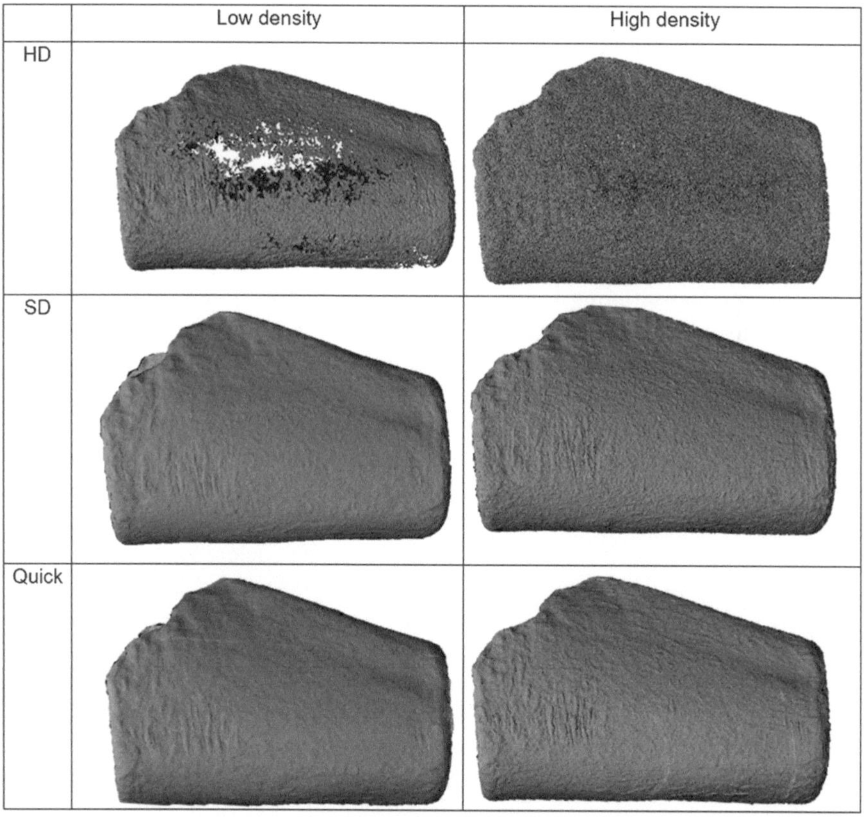

Fig. 8.5 Comparison of the influence of resolution and point density on the quality of a model – in this case a carnivore modified bone. Key: HD, low 29,000 and high 268,000 points; SD, low 4200 and high 16,000 points; quick, low 1200 and high 3300 points

when scanning even relatively simple objects – too low a point cloud density and voids or hole appear. Too high a resolution, and computational artefacts may be introduced.

8.5 Discussion

At present there have been few published studies which seek to validate the quality of the scanning process and post-processing outcomes, and much more work is required to produce industry-wide standards for the use of surface capture technologies. We hope that it will eventually be possible to produce a set of standards and protocols like those used in forensic photography. It is also important to stress the requirement for formal training before practitioners engage in 3D capture and modelling. Scanning and model production can be a complex and time-consuming process, and there is an implicit cost-benefit trade-off that must be understood before 3D capture methods can be effectively introduced into common forensic practice. With finite budgets within police forces and forensic providers, the cost of 3D capture technology must be justified – especially as some capture modalities costing in the tens of thousands of dollars. With 'gold standard' micro-computed tomography costing more than half a million dollars on average, it is unlikely that this modality will be adopted in general practice and research anytime soon, though it does present the best available option for the establishment of standards against which all other methods may be compared.

All is not doom and gloom however. It is clear from this paper and other published studies that differing technologies and modalities of surface capture have their own distinct advantages and disadvantages, with relative cost being a major factor – but even relatively cheap capture technologies can provide significant advantages if adopted and used effectively and with proper training in surface capture and model creation. We note that there is an upward trend in the utilisation of 3D scanning technologies and sharing platforms across forensic science, and this trend appears to be driven by improvements and commercialisation of 3D capture technology, leading to a fall in prices and wider availability. This may even see bolt-on devices, such as the IO Structure for iPad (https://structure.io) achieving wide acceptance and use. This growth in the range of available technology is being led by the expanding market in home uses for 3D modelling and printing; however, it makes selecting the most appropriate capture method for forensic practice more complex. On this basis, it can be cogently argued that photogrammetry provides an all-round best-fit approach when considering user friendliness and ease of use, relative cost, and its capability of recording texture which may be important in taphonomic studies. However, photogrammetry does suffer from issues of dimensionality (a lack of ability to embed absolute scaling of models in certain photogrammetric software packages), depth resolution and separation of fine surface topography, and

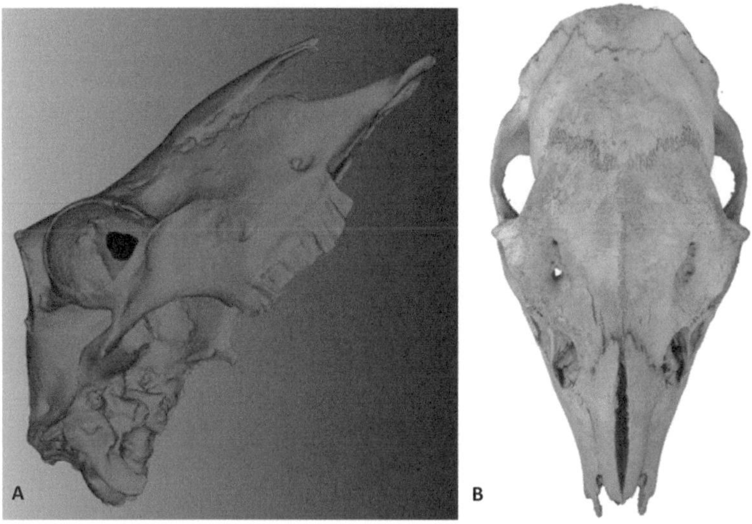

Fig. 8.6 Comparison of 3D scan quality using (**a**) Artec Eva and (**b**) photogrammetry. Note the lack of fine surface geometry in A (such as patination, crazing, and cranial sutures) which is recorded in B

false-surface production in translucent subjects. In many cases (particularly site-based studies), it may be applicable to use a multimodal method combining the precise geometry capture associated with laser scanning with the texture-capture abilities of photogrammetry; both are combined in devices such as the NextEngine scanner, but in fact they may be better implemented in practice from two different high-specification devices, rather than one suboptimal device (Fig. 8.6).

It is fair to say that there is much work that is needed before 3D scanning technology is fully integrated into forensic taphonomic analyses. At present, the technology exists to capture, visualise, and share surface shape models for the purposes of research or casework. Surface scanning should be undertaken with the explicit aim of digitising, comparing, reconstructing, and sharing objects, thus providing a quantitative framework for the analysis of taphonomic alterations to objects and sites. However, such aims require standardisation of technology and methodologies employed sufficient to fulfil the requirements of laboratory accreditation and evidential admissibility in court. There is still a major gap between user and courtroom as to the full capability of 3D scanning; this may require protocols to fully standardise 3D capture technologies within the world of forensic practice, with the production and implementation of protocols and known error margins associated with the use of such models. Theoretically it should be possible to construct a shape space dataset of known experimentally produced taphonomic traces on bone (abrasion, tool and cut marks, blunt and ballistic trauma, ichnotraces, etc.) such that the data could be readily shared between taphonomists and analysed or examined with

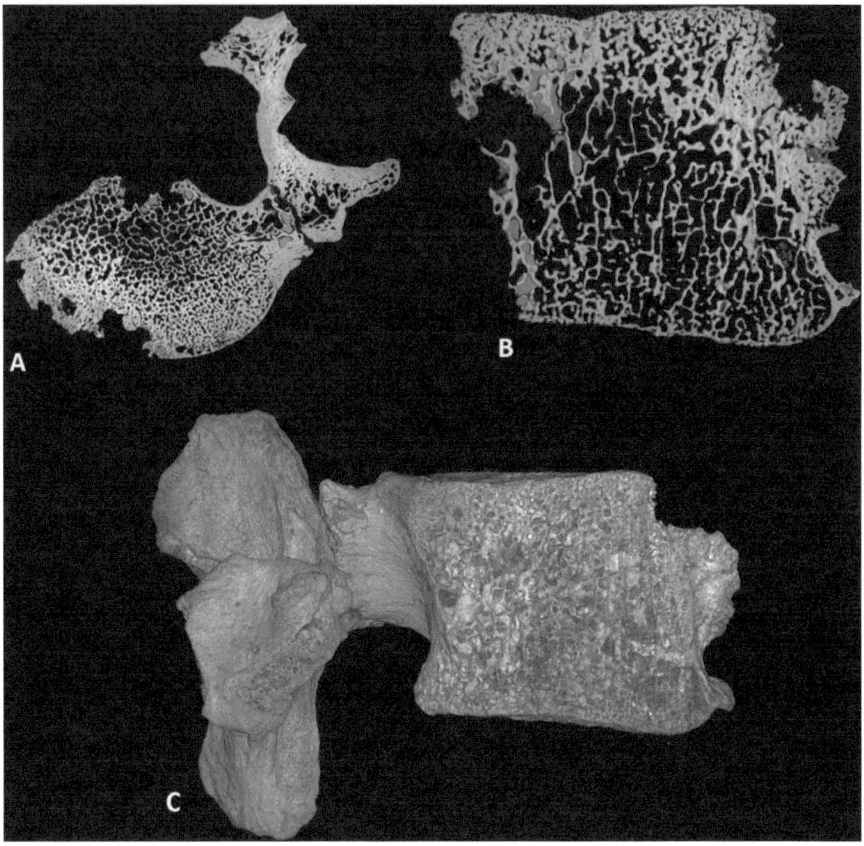

Fig. 8.7 The 'gold standard' in recording taphonomic bone traces. (**a**) and (**b**) show orthosclice views through a lumbar vertebra affected by invertebrate boring and surface modification (ichnotraces). (**c**) shows a surface-rendered model based on reconstruction of the tomographic image volume. Reconstructions undertaken in Avizo Amira 5.4. Specimen was imaged using a Nikon Metrology XTH 225/320 LC dual source industrial CT system and captured at a resolution of 33 µm. (Adapted from Odes et al. 2017 with permission)

a common objective methodology. As noted above, micro-computed tomography represents the best available option for the establishment of standards against which all other methods may be compared, particularly when such a modality allows for the imaging and modelling of both internal and external structures and taphonomic alterations (see Fig. 8.7) to both.

Chapter 9
Human Remains in Marine Environments: Challenges and Future Developments

Agathe Ribéreau-Gayon, Carolyn Rando, and Ruth M. Morgan

Abstract Human remains immersed in marine environments present specific challenges at various steps of the forensic investigation process, including the detection, recovery, analysis, and interpretation of the remains. These challenges need to be clearly identified in order to implement suitable solutions to maximise the chances of recovery of the remains and reduce the risk of uncertainty in the forensic reconstructions, including the biological profile of the deceased. This chapter addresses these issues by following the sequential steps within a forensic investigation in marine environments, focussing on the taphonomic factors and patterns that can impact accurate reconstructions of the taphonomic sequence, from the time of recovery of the remains back to the time of death.

Keywords Aquatic forensics · Forensic reconstructions · Marine taphonomy · Underwater investigation

A. Ribéreau-Gayon (✉)
Department of Security and Crime Science, University College London, London, UK

Centre for the Forensic Sciences, University College London, London, UK

Institute of Archaeology, University College London, London, UK
e-mail: agathe.ribereau-gayon.13@ucl.ac.uk

C. Rando
Institute of Archaeology, University College London, London, UK
e-mail: c.rando@ucl.ac.uk

R. M. Morgan
Department of Security and Crime Science, University College London, London, UK

Centre for the Forensic Sciences, University College London, London, UK
e-mail: ruth.morgan@ucl.ac.uk

P. M. Barone, W. J. M. Groen (eds.), *Multidisciplinary Approaches to Forensic Archaeology*, Soil Forensics, https://doi.org/10.1007/978-3-319-94397-8_9

9.1 Introduction

Water, in liquid and frozen forms, covers about 75% of the Earth's surface, with the vast majority (96.5%) being the oceans and seas (Graham et al. 2010). In this respect, aquatic environments and specifically marine environments represent areas of potential forensic interest as significant environments in which a forensic reconstruction of events can be valuable in crime detection. There are approximately 360,000 deaths by drowning worldwide each year (World Health Organization 2017), and approximately 5000 migrant deaths were reported in 2016 in the Mediterranean Sea (International Organization for Migration 2017; The United Nations High Commissioner for Refugees 2016). Despite the increasing number of fatalities in aquatic environments, the methods used to establish the postmortem submersion interval (PMSI), the cause and manner of death, and positive identification of the remains have been argued to lack robustness, transparency, and accuracy (Petrik et al. 2004; Anderson and Bell 2014, 2017; Dickson et al. 2011; Barrios and Wolff 2011; Ribéreau-Gayon et al. 2017; Ellingham et al. 2017). There is, therefore, a significant gap in the understanding of how human remains decompose in aquatic environments (the field of aquatic taphonomy) and how this can impact the ability to not only efficiently search and recover submerged remains but also produce accurate forensic reconstructions (Ribéreau-Gayon et al. 2017). These issues can lead to uncertainty in the presentation of taphonomic evidence to court. This is particularly noticeable in investigations that involve marine environments as forensic marine archaeology and anthropology are poorly developed in comparison to terrestrial environments where most of the casework and published research to date have been carried out (Megyesi et al. 2005; Keough et al. 2017; Simmons et al. 2010a, b; Maile et al. 2017; Vass 2011). This can be explained in part by the technical difficulties of conducting experimental research in open and multivariate marine environments. Indeed, the proportions of surface of deep seas, *sensu lato*, currently known or estimated range between about <0.0001% and 10% pending on the areas considered (Ramirez-Llodra et al. 2010). Nevertheless, further studies are needed to generate empirical data that can contribute towards the development of a robust evidence base in the field of forensic marine archaeology and anthropology.

This chapter provides an overview of the current issues in forensic marine investigations, focussing on the detection, recovery, and identification phases of the investigation process (Morgan 2017a) and discusses the key role of evidence-based research in addressing these issues. The chapter focusses on the taphonomic patterns that were observed in a forensic dataset, currently studied by the authors, and their likely origin (the taphonomic factors) that resonate with the published literature. Other aspects of marine taphonomy, including some patterns of geo- and bioturbation, will not be explored here as they are already addressed in detail in the existing published literature (Higgs and Pokines 2013; Bell 2011). The issues addressed in this chapter will be illustrated through casework examples from two airplane accidents in marine environments: the Air France crash that occurred on 1 June 2009 in the Atlantic Ocean not far from Brazil and the Yemenia crash that occurred on 29 June 2009 in the Indian Ocean, in the Comoros Islands. The human remains analysed here were recovered at 3900–4000 m deep for the Air France crash and 1200 m deep for the Yemenia, with maximum PMSIs of 24 months and 59 days, respectively.

The chapter is organised to follow the sequential steps within a forensic investigation in marine environments, from the detection and recovery of the remains (2) to their subsequent analysis and interpretation (3). The main perspectives within the field of marine forensic taphonomy are then presented (4). It is important to note that the taphonomic factors that have an impact on the ability to accurately analyse and interpret human remains can also have an impact on the ability to effectively detect and recover them in the earlier phases of the investigation. The pertinent taphonomic factors will thus be specifically discussed in Sect. 9.3.

9.2 The Issues Associated with the Detection and Recovery of Human Remains in Marine Environments

Efficient management of the remains in the first phases of the forensic investigation is essential to enable accurate forensic reconstructions given that each step of the investigation process is reliant on the prior steps (Inman and Rudin 2002). Thus, any information or items missed or incorporated in the first phases of an investigation have the potential to cascade or even snowball into the subsequent phases of the investigation which can impact the interpretation of the evidence and its presentation to the investigators (Dror et al. 2017). This can have a significant impact upon the eventual outcome of a case (Nakhaeizadeh et al. 2018). Aquatic environments, including marine environments, present specific challenges that make detection and recovery of human remains particularly delicate (Boyle et al. 1997; Sorg et al. 1997; Ebbesmeyer and Haglund 2002; Heaton et al. 2010; Simmons and Heaton 2013) and sometimes more so than on land. The challenges of detecting and recovering human remains mostly lie in the vast surfaces and great depths encountered in marine scenarios. The properties and composition of the water also play a critical role in the degradation or preservation of the remains and thus on the ability to retrieve them.

9.2.1 Water Properties and Composition

Three main types of factor interact during the decomposition process in marine environments: biological, chemical, and physical (Simmons and Heaton 2013). Variables contingent upon these factors are presented in Table 9.1. Specific elements must be considered when examining human remains found in water. The type of water (e.g. freshwater or saltwater, still or flowing) is most important (Lawler 1992). Publications further stress the need to distinguish, within saltwater, between shallow, deep water, coastal water, and offshore water environments (Anderson and Hobischak 2004).

Table 9.1 Factors – exogenous and endogenous – that impact the decomposition patterns and rates of human remains in marine environments (adapted from Lawler 1992; Haglund and Sorg 2001; Dumser and Türkay 2008; Simmons and Heaton 2013)

Type of factors	Factors	
Exogenous	Biological	Fauna
		Flora
		Fungi
		Algae
		Microorganisms (including bacteria and plankton)
	Chemical	Salinity
		Oxygen level
		Water composition
		pH
		Density
		Turbidity
		Sea floor composition
	Physical	Temperature
		Depth
		Light
		Geographic location
		Currents
		Tidal action
		Pressure
Endogenous		Movements of the remains
		BMI
		Clothing

Indeed, water depth directly influences the major factors in the decomposition process, such as light, temperature, oxygen, currents, and substrate (Lawler 1992; Haglund and Sorg 2001; Dumser and Türkay 2008). Accordingly, the findings on the human remains and their preservation in a marine context depend on a number of exogenous factors, such as environmental conditions (Kahana et al. 1999; Dumser and Türkay 2008; Anderson and Bell 2014). Immersion in seawater can modify or even conceal the traces of trauma, whether peri- or postmortem, because of the erosion of the edges of the lesion, loss of body parts, and acceleration of the decomposition process (Ribéreau-Gayon et al. 2017; Stock et al. 2017). Additionally, the action of marine currents, waves, and tides tumbles the bones, leading to rounding on their edges (Higgs and Pokines 2013). Poor visibility conditions, including due to the turbidity of the water, can be an extra challenge to efficiently detect and recover immersed human remains. These various challenges call for specific training, resources, and equipment that are adapted to the marine environment. However, current knowledge on the precise role of the factors linked with water properties and composition, for both isolated and aggregated sets of variables, is patchy. This is problematic given that the type of aquatic ecosystem directly impacts the movements of human remains in water, which drive the decomposition process.

9.2.2 Movements of Human Remains in Marine Environments

It has been observed that there is a variability in the patterns of movements of human remains that are immersed in marine environments. Some float, while others sink immediately (Heaton et al. 2010). It is important to consider that in marine environments the remains can move in all the three dimensions, in contrast to terrestrial environments (Simmons and Heaton 2013). The taphonomic factors involved in the patterns of movements of the remains are mentioned in Table 9.1. Whether the cadaver or remains sank or floated after death/immersion in water impacts the amount and types of postmortem changes induced by aquatic fauna. Easily accessible remains, such as floating cadavers, are likely to attract a great variety of aquatic and terrestrial animals, such as insects and birds (Petrik et al. 2004). Thus, movements of cadavers in water may play a decisive role in decomposition rates (Petrik et al. 2004). In addition to the intrinsic movements of the remains due to the decomposition process, natural (e.g. currents and tides) and recreational activities (e.g. boating, nautical sports, or fishing) can impact the movements of the remains, sometimes leading to drift patterns. These scenarios cause extreme scattering of the remains which makes their detection very challenging (Ebbesmeyer and Haglund 1994; Mateus et al. 2015; Carniel et al. 2002; Commission d'Enquête 2013).

Case Study

Drift patterns of human remains on very long distances were observed in the case of the Yemenia airplane crash. A week after the crash, debris of the plane wreckage and human remains were retrieved in Tanzania up to about 900 km and 530 km from the crash site, respectively. The drift patterns and distances were reconstructed by the French Naval Hydrographic and Oceanographic Service (SHOM) and are presented in Fig. 9.1 (Commission d'Enquête 2013). This study determined that the long drift distances over a short period of time were due to the strong oceanic currents in the Mozambique Channel. This casework example demonstrates the importance of collaborative work between oceanographers, marine biologists, and forensic anthropologists to understand the drift patterns of human remains in marine environments in order to develop predictive models to help prioritise the search and maximise the chances of recovery of the remains. Some retrospective and experimental studies focussed on reconstructing the drift of human remains in marine environments have suggested some models for these movement patterns (Sherman and Nordstrom 1994; Pampın and Rodrıguez 2001; Mateus et al. 2013, 2015; Duijst and Passier 2016). However, the movements of human remains in water are still poorly understood, highlighting the need for further experimental research.

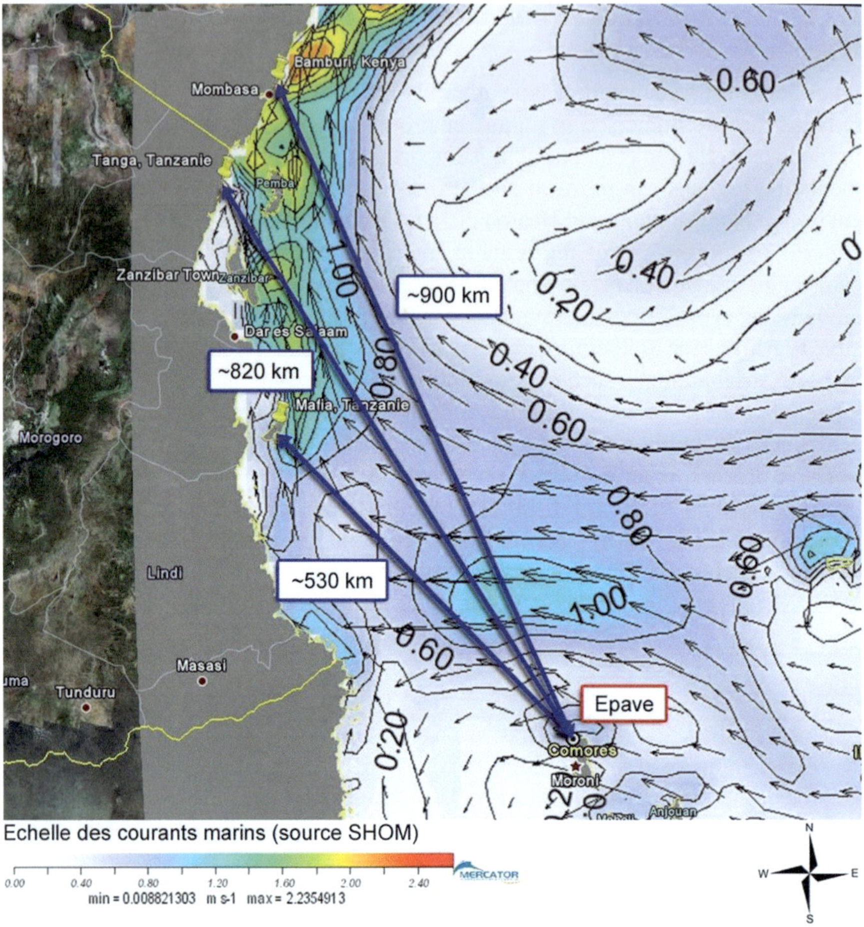

Fig. 9.1 Reconstruction of the drift of the debris and human remains from the Yemenia airplane crash (29th June 2009) with regard to the ocean currents (Commission d'Enquête 2013). Legend "Epave" = wreck, "Comores" = Comoros, "Echelle des courants marins" = scale of marine currents

The depth of the water is a really important factor to consider as shallow and deep marine waters require different investigative approaches. Detecting human remains submerged in deep seas, from about 200 m deep (Gage and Tyler 1991; Ramirez-Llodra et al. 2010), requires the deployment of specific means such as ground-penetrating radar (GPR) (Ruffell 2005) and side-scan sonar (Healy et al. 2015), including placed on board autonomous underwater vehicles (AUVs) (Purcell et al. 2011) and remotely operated vehicles (ROVs) (Commission d'Enquête 2013; Murphy et al. 2011). Both AUVs and ROVs are unmanned underwater vehicles with a noticeable difference; while AUVs operate independently with no connecting cables, ROVs are connected and operated from the deck of a ship at the surface of the water (Purcell et al. 2011; Murphy et al. 2011). The ROVs can be equipped with

Fig. 9.2 Recovery process of the human remains from the Air France and Yemenia airplane crashes (1st and 29th June 2009) in oceanic environments. (**a**) Remotely operated vehicle (ROV) Remora 6000 by Phoenix International that was deployed to retrieve the remains of the victims of the Yemenia crash (29th June 2009). www.bea.aero.com. (**b–d**) Collection of human remains on the deck of a ship at the surface of the water, Air France crash (1st June 2009). IRCGN

a light (as depths greater than 1000 m tend to be dark environments with no natural light (National Oceanic and Atmospheric Administration 2017)) and a video camera which enables to monitor the surroundings of the ROVs in real time (Fig. 9.2a). The ROVs are also equipped with an articulated arm that extends with metallic gripping claws to enable the collection of evidence. This approach has been successfully used in a number of high-profile forensic cases that involved human remains immersed at great depths, including the Yemenia and Air France crashes in June 2009 (1200 and 3900/4000 m deep, respectively) (Fig. 9.2b–d) (Commission d'Enquête 2013; SHOM 2009).

The study of the Air France and Yemenia airplane crashes in different marine environments identified that the recovery process of human remains that have been immersed for weeks leads to an immediate and irreversible degradation of the soft tissues (Ribéreau-Gayon et al. 2017). This is due to (i) the motion made when the

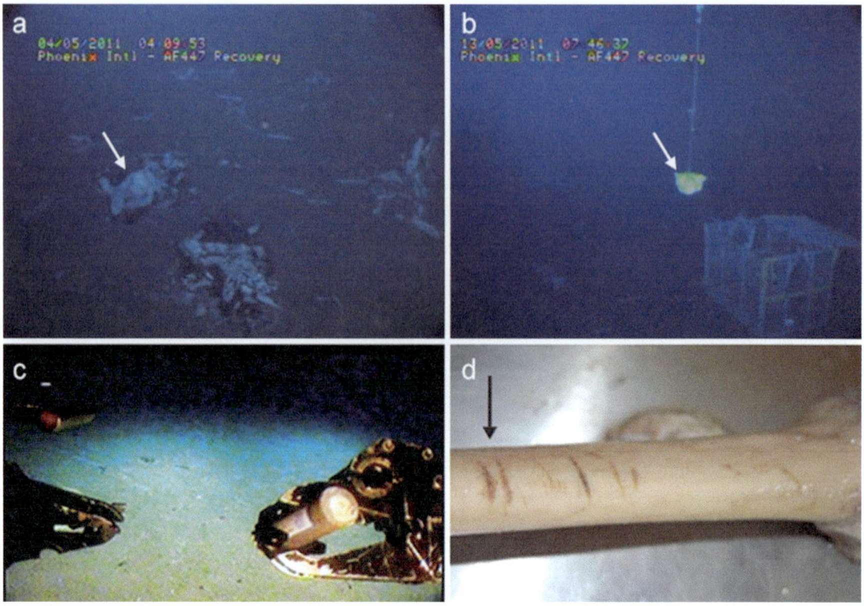

Fig. 9.3 The issues of detecting and recovering the human remains and debris from the Air France and Yemenia airplane crashes (1st and 29th June 2009) at great depths. (**a**) Detection of the human remains (arrow) and debris from the crash lying against the ocean floor, Air France crash (1st June 2009). IRCGN. (**b**) The human remains are lifted (arrow) and placed into a basket by the ROV to enable their subsequent collection at the surface of the water, Air France crash (1st June 2009). IRCGN. (**c**, **d**) The ROV's metallic gripping claws (**b**) likely left parallel striae (arrow) on the external surface of some bones (**d**), Yemenia crash (29th June 2009). IRCGN

remains are pulled up to the surface of the water and (ii) a brutal change in the composition of the environment surrounding the remains, something we aim at investigating further in future work. Therefore, collecting the remains and bringing them up to the surface of the water (Fig. 9.3a, b), while maintaining their integrity, represents additional challenges that must be taken into consideration when undertaking a forensic marine investigation. For instance, the mechanic recovery process may cause damage to the remains and thus introduce potential uncertainty at the analytical and interpretation phases in the forensic investigation.

Case Study

It was identified in the case of the Yemenia crash (Ribéreau-Gayon 2014) that the metallic articulated claws of the ROVs can leave parallel striae on the external surfaces of the bones (Fig. 9.3c, d). These patterns presented some macroscopic similarities with other peri- or postmortem patterns such as grooves caused by the teeth of large sharks. Thus, documentation on the environment of decomposition of the remains and the recovery process is critical to enable the understanding of the origin of the patterns that may be subsequently observed in a laboratory.

9.3 The Issues Associated with the Analysis and Interpretation of Human Remains Recovered from Marine Environments

The identification of the deceased and the diagnosis of the circumstances that led to their death are the ultimate aims of the forensic investigation. However, accurate postmortem conclusions in marine investigations can be complicated by the impact of various marine taphonomic parameters upon the preservation of the remains, as highlighted in Table 9.2 (Simmons and Heaton 2013; Schuliar et al. 2014; Ribéreau-Gayon et al. 2017).

9.3.1 The Taphonomic Patterns in Marine Environments

Human remains immersed in water go through a number of stages of decomposition over time (Heaton et al. 2010; van Daalen et al. 2017). Some of these decomposition stages are specific to aquatic environments in their physical expressions (taphonomic patterns) and rates (van Daalen et al. 2017). Accurate analysis of human remains is dependent upon the level of preservation, which is affected by parameters that are often

Table 9.2 The taphonomic patterns in human remains immersed in marine environments associated with some of the most commonly reported factors responsible for these patterns

Taphonomic patterns	Taphonomic factors
Maceration and skin slippage	Immersion in water
Differential decomposition	Immersion in water
	Movements of the remains
	Currents and tidal action
	Water properties and composition
	Terrestrial environment
Adipocere	Immersion in water
	Water properties and composition
	Body coverings
Disarticulation	Immersion in water
	Time (PMSI)
	Water temperature
	Body coverings
	Faunaturbation
	Floraturbation
	Microorganisms
	Geoturbation
	Recovery process (anthroturbation)
	Water properties and composition
Trauma	Faunaturbation
	Cause and manner of death

dependent upon specific properties of the human remains themselves (endogenous factors), such as the presence of clothing or trauma, as well as their environment of decomposition (exogenous factors), such as faunal and floral activity (Simmons and Heaton 2013). Some of these factors are interdependent and thus impact the taphonomy of the human remains in a complex manner. Taphonomic changes can conceal features that are valuable to the forensic expert, thus making accurate reconstruction of the sequence of events from the time of death to the time of recovery of the remains (taphonomic sequence) difficult (Ribéreau-Gayon et al. 2017; Ribéreau-Gayon et al. 2018).

Maceration and Skin Slippage

Maceration corresponds to waterlogged superficial layers of the skin – epidermis – and is typically observed in the early stages of immersion in water, including saltwater (Farrugia and Ludes 2011; Simmons and Heaton 2013). In the early stages, the skin looks pale and wrinkled. In more advanced decomposition stages, the waterlogged epidermis skin starts slipping off the underlying tissues (skin slippage), often in the form of patches (Hobischak 1997; Heaton et al. 2010; Humphreys et al. 2013; van Daalen et al. 2017), as shown in Fig. 9.4a. The most advanced expression of this phenomenon in the hands and feet takes the form of "degloving" where the entire superficial skin on the region is separated from the underlying tissues, such as fat, tendons, ligaments, and muscles. This particular feature is referred to as "washerwoman hand" in the hands, as shown in Fig. 9.4b, c (Heaton et al. 2010; van Daalen et al. 2017). However, no tool currently exists to precisely quantify the amount of skin slippage, either its depth or its extent over the cadaver. Features of maceration are helpful indicators of immersion, although they cannot be used to predict the type of water that the human remains were in, such as freshwater, seawater, flowing water, or still water. In order to identify the environment of death and/or decomposition, information needs to be collected from complementary sources.

Differential Decomposition

In marine environments, the remains can be subject to various conditions and movements leading to, for example, one part being exposed above the water line and the other being submerged (Haglund 1993), as shown in Fig. 9.5. These circumstances cause a combination of decomposition patterns (differential decomposition). For instance, mummification and adipocere can coexist on a single immersed cadaver or body part if they are immersed in water and exposed to direct sunlight and hot ambient temperatures all at once, causing exposed soft tissues (e.g. face, thorax, and abdomen) to dry and potentially mummify (Haglund 1993). This phenomenon of differential decomposition has also been reported on land (Shean et al. 1993; Cockle and Bell 2015).

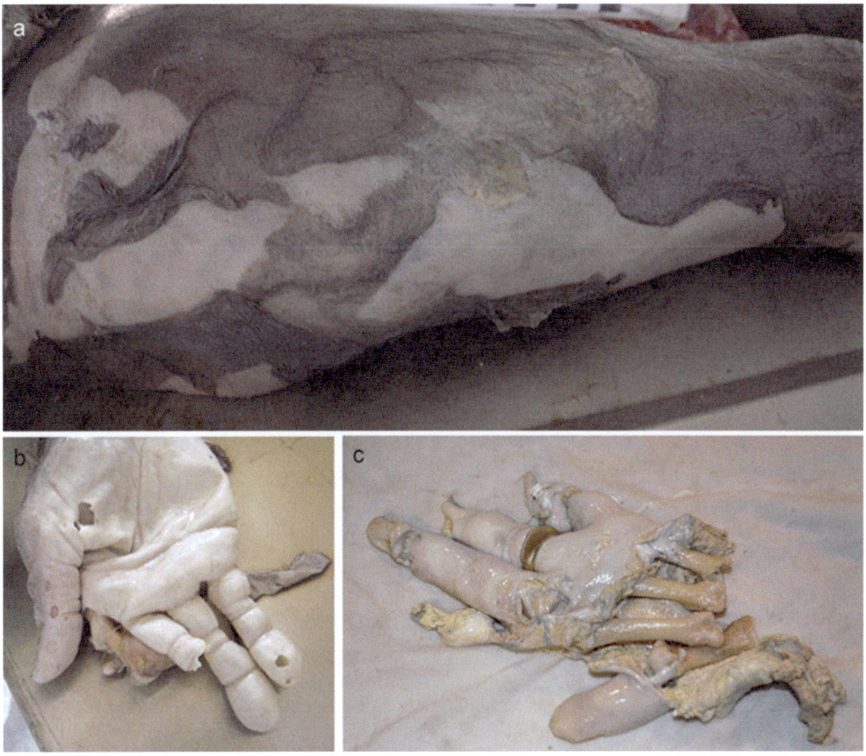

Fig. 9.4 Skin slippage and degloving in the victims of the Air France and Yemenia airplane crashes (1st and 29th June 2009). IRCGN (**a**) Early skin slippage on the leg in the form of patches (Yemenia crash, 29th June 2009). (**b**) Advanced skin slippage on the hand in the form of degloving ("washerwoman hand"), resulting in the exposure of the underlying tissues (**c**) (Air France crash, 1st June 2009)

Adipocere

Adipocere is known as the "fat of graveyards" (Wetherill 1860) and is one of the most prominent features encountered by forensic pathologists and anthropologists in humid environments, particularly at aquatic forensic scenes (Fig. 9.6). The occurrence of adipocere, including in saltwater, has been extensively studied (Kahana et al. 1999; O'Brien and Kuehner 2007; Forbes et al. 2002, 2005, 2011; Ubelaker and Zarenko 2011; Widya et al. 2012; Notter and Stuart 2012; Stuart et al. 2005, 2016; Forbes and Stuart 2004; Ueland et al. 2014; Stuart and Ueland 2017a). Adipocere leads to a transformation of the tissues, including the skin and adipose tissues (fat), into a waxy texture that forms a layer which preserves the underlying tissues such as the internal organs and the bones (Fig. 9.6). Different morphological descriptions of adipocere exist in the literature. It is thought that adipocere has two distinct stages: fresh and advanced. The fresh stages correspond to a soft, greasy appearance, while the later stages are characterised by a hard and brittle texture

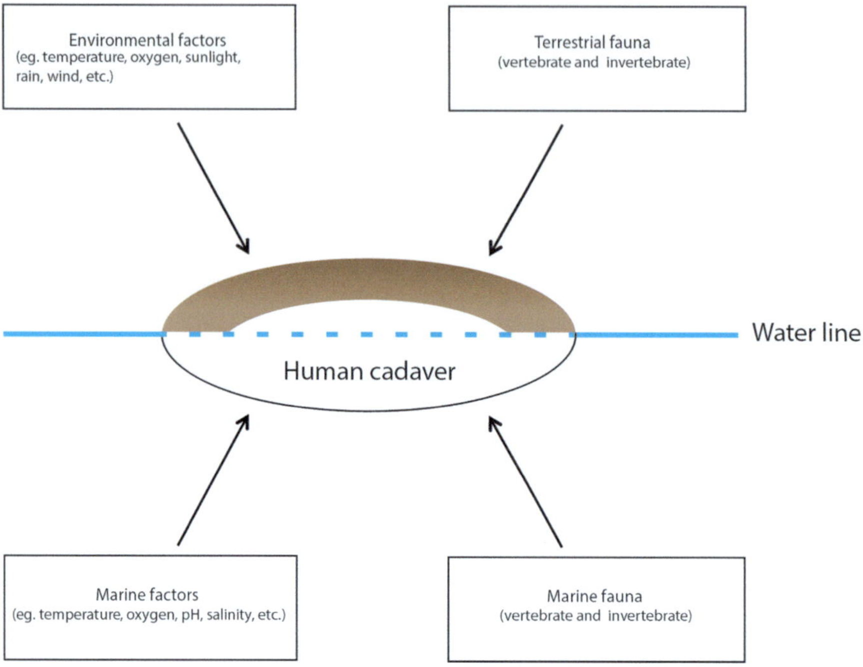

Fig. 9.5 Schematic representation of the phenomenon of differential decomposition at sea

(Ubelaker and Zarenko 2011). Not only does the texture of adipocere change over time, but also the colour changes, from yellow in the fresh stages to white in the advanced stages. Despite the various macroscopic descriptions of adipocere, there is a reasonable consensus regarding the origin of adipocere as a by-product of decomposition formed by the hydrolysis and hydrogenation of adipose tissues (Kahana et al. 1999).

The phenomenon of adipocere particularly interests taphonomists because it fosters a lasting preservation of the remains which is sometimes sufficient to establish a biological profile of the deceased that can lead to a positive identification and a determination of the cause and manner of death (Kahana et al. 1999; Ubelaker and Zarenko 2011). However, because the presence of adipocere 'stops' the decomposition process by preserving the tissue underneath, it may introduce uncertainty in the estimation of the decomposition stage of the cadaver, possibly leading to erroneous PMSI estimates (Heaton et al. 2010), usually in the form of underestimating the PMSI given the overall preservation of the remains. The rates of formation of adipocere are accelerated by the presence of clothing and wrapping (Notter and Stuart 2012; Stuart et al. 2016), which will be discussed in more detail in the section "The Presence of Garments and Body Coverings".

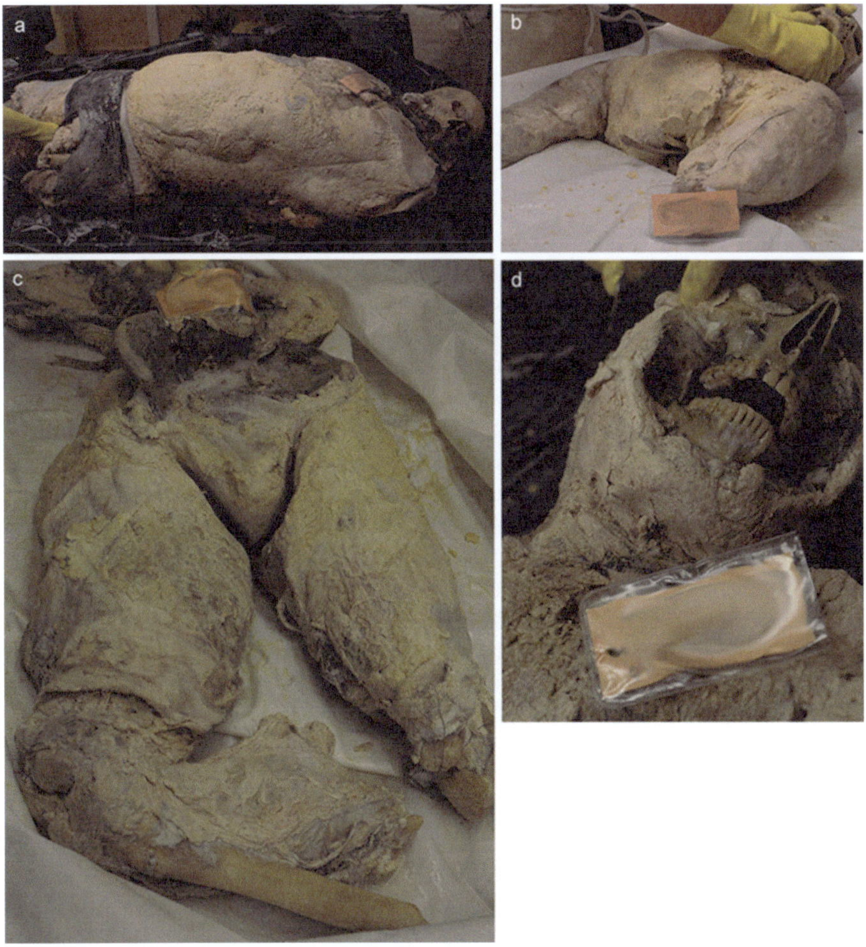

Fig. 9.6 Formation of adipocere (saponification) on victims of the Air France airplane crash (1st June 2009) after 2 years of immersion in the Atlantic Ocean at a depth of 3900/4000 m. IRCGN Adipocere was observed on various body parts: (**a**) the trunk, (**b**) the upper thorax and the proximal upper limbs, (**c**) the pelvis and the proximal lower limbs, as well as (**d**) the neck and the cheeks

The Sequence of Decomposition and Disarticulation, and the Decomposition Rates

In water, the sequence of disarticulation follows an outside to inside pattern, commencing at the extremities of the limbs (e.g. hand–wrist and foot–ankle regions), followed by the head–neck region (the mandible being first) and then the lower legs, forearms, and upper arms (Haglund 1993). The trunk, including the pelvic girdle with the femora still articulated, is the body region that is preserved the longest (Haglund 1993). This sequence is time-related, thus providing an indication of PMSI. However, to date the sequence of disarticulation of human cadavers in aquatic environments has mainly been described by Haglund and Sorg (Haglund 1993; Haglund and Sorg 1996), but these findings still require empirical validation. This is particularly important as recent experimental studies have shown diversity in the onset of complete skeletonisation in aquatic environments, ranging from a few days to several months (Anderson and Bell 2016). There is also currently a lack of consensus regarding the rates of decomposition in freshwater compared to saltwater environments, some studies showing that decomposition in saltwater occurs faster than in freshwater (Buchan and Anderson 2001; Dumser and Türkay 2008), while others demonstrate the opposite finding (Ayers 2010; Introna et al. 2013). It appears that these discrepancies between results can be explained by the presence or absence of fauna activity. Specifically, when present in saltwater environments, necrophagous fauna accelerates the decomposition process compared to freshwater environments. For example, in marine environments, fauna can cause complete skeletonisation of a cadaver within 1 month, in comparison to a minimum of 3 months in lacustrine environments (Buchan and Anderson 2001; Dumser and Türkay 2008). However, when postmortem fauna activity does not occur (e.g. anaerobic conditions, enclosed scenes, or intentionally prevented in experimental studies), decomposition in saltwater actually takes longer than in freshwater (Ayers 2010; Introna et al. 2013). Therefore, while significant advances have been made in our understanding of these processes, it is clear that further studies are needed to better understand the impact of multiple, complex sets of variables on the decomposition process in marine environments, such as the combination of water composition and fauna activity.

Approaches to Assess Decomposition and Estimate the PMSI

The sequence of taphonomic patterns can be used to estimate the PMSI of human remains in marine environments. To date, one method has specifically been developed based on these patterns, the van Daalen et al. (2017) method. The method builds upon the literature available in the field of terrestrial taphonomy (Galloway 1997; Megyesi et al. 2005) and consists of allocating a numeric score to the head–neck, trunk, and limbs according to the taphonomic patterns observed, such as skin slippage, hair and nail shedding, or bloating. The sum of these three decomposition scores provides a total aquatic decomposition score (TADS) which can then help estimating the PMSI. Evaluating the suitability of this method for a range of different marine environments, including in deep oceanic conditions, such as those encountered in the Air France and Yemenia cases, would help strengthen the PMSI estimates in forensic marine investigations.

Furthermore, the potential for other indicators, such as algal (Keiper and Casamatta 2001; Casamatta and Verb 2000; Zimmerman and Wallace 2008), microbiological (Tarr 1954; Haefner et al. 2004; Benbow et al. 2015; Dickson et al. 2011), and arthropod (Anderson and Bell 2014, 2016; Magni et al. 2015) communities, to help estimating the PMSI in marine environments is currently being explored. Additional experimental studies are needed to determine the accuracy and applicability of these approaches as well as evaluating their capacity to assist with establishing the PMSI when applied in an integrated and corroborative manner.

9.3.2 The Taphonomic Factors in Marine Environments

Geoturbation

Contact with the seabed, sometimes comprised of silt, sand, or sharp rocky surfaces, can cause distinctive patterns of abrasion and staining on the external surfaces of the bones (Pokines and Higgs 2015; Higgs and Pokines 2013; Pokines 2018).

Case Study

This combination of patterns was observed in the case of a skull recovered from the Air France airplane crash on the oceanic bed at a depth of approximately 4000 m in the Atlantic Ocean. The frontal bone exhibited unusual linear scraping marks, as shown in Fig. 9.7. These marks were probably due to abrasion by the sediment (Pokines and Higgs 2015), as the face was likely lying against the seabed, as indicated by the dark colouration of the bones of the face (Pokines 2018) in contrast with the rest of the skull where the external layer of the bone has been destroyed, perhaps due to faunal activity. However, it cannot be ruled out that the linear marks on the frontal may also have been caused by the teeth of large sharks scavenging on the remains. The taphonomic impact of these large scavengers is discussed in more detail in the section "Faunaturbation – Larger Fauna".

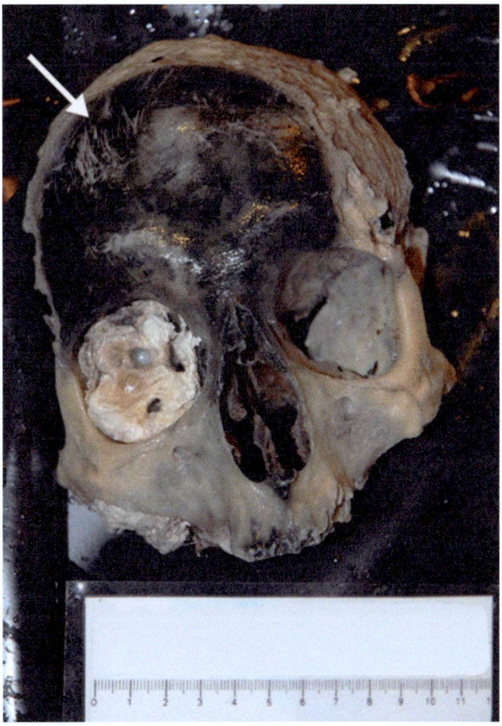

Fig. 9.7 Patterns of staining and differential decomposition on a human skull, associated with intriguing scraping marks on the frontal bone (arrow), Air France crash (1st June 2009). IRCGN

The impact of fauna (faunaturbation), whether small or larger, on taphonomy of human remains needs to be understood by the forensic investigators because micro-organisms and fauna are present in a wide range of marine environments. The impact of flora (floraturbation), algae, plankton, and other microorganisms is, however, yet to be established and is currently an area in need of systematic empirical study to assess the degree to which these processes are factors in the taphonomic process in marine environments.

Faunaturbation

Faunaturbation to human remains in marine environments can be categorised into small and larger fauna.

Small Fauna

Macro-boring and grazing caused by *Osedax* ("bone-eating worms") (Johnson et al. 2010; Alfaro-Lucas et al. 2017; Vrijenhoek et al. 2009; Aronson et al. 2016; Smith et al. 2015; Tresguerres et al. 2013) and *Rubyspira osteovora* ("bone-eating snails") (Johnson et al. 2010; Aronson et al. 2016; Smith et al. 2015), two species of small fauna, have been reported a number of times on bones of large marine mammals, such as whales. Experimental studies on whale carcasses showed that the morphology of the borings caused by the *Osedax* is variable and highly dependent on the type of bone (Higgs et al. 2011a, b; Smith et al. 2015). These patterns of macro-boring are clearly different from that of micro-boring (or tunnelling) that were observed in the macrostructure of bones in some oceanic ecosystems (Bell et al. 1996; Bell 2011; Bell and Anderson 2016; Pesquero et al. 2017). Patterns of micro-tunnelling consist in superficial tunnels (50–100 microns deep) in the cortical bone with a diameter of 5–7 μ (Bell and Anderson 2016). These taphonomic patterns are likely caused by endoliths, but the circumstances under which they occur are not fully understood to date.

Case Study
Several human bones recovered from the Air France crash after 2 years of immersion showed macro-borings that shared some common characteristics with the patterns described in the marine biology literature (Fig. 9.8). However, as no bone-eating worms or snail borings have yet been reported on human bones, accurate identification of the patterns observed in this case is yet to be established. This case study demonstrates that the impact of marine microfauna in the breakdown of human remains requires further study.

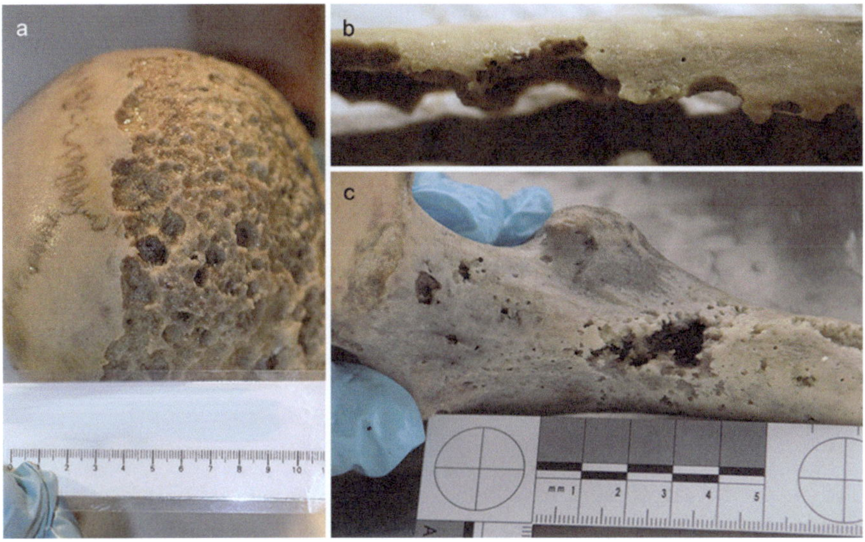

Fig. 9.8 Unusual patterns of boring and grazing on a skull (**a**) and long bones (**b** and **c**) recovered from the Atlantic Ocean, Air France crash (1st June 2009). IRCGN

Faunaturbation on decomposing remains may mimic perimortem trauma, thereby introducing uncertainty in the identification of the trauma and the forensic reconstructions (Byard et al. 2002; Ribéreau-Gayon et al. 2017). This outlines the vital importance for forensic experts to have a solid understanding of the postmortem patterns caused by marine fauna on human remains to enable reliable postmortem conclusions. The literature outlines the importance of two categories of small invertebrates in the decomposition of human remains: arthropods and molluscs. The published literature on the role of arthropods in a forensic context focusses on empirical research conducted by multidisciplinary teams (e.g. collaborations between entomologists and forensic anthropologists) (Anderson and Bell 2014, 2017). Among forensically relevant arthropods, crustaceans such as crabs, crawfish, and barnacles have been particularly studied in the published literature (Anderson and Bell 2014, 2017; Magni et al. 2015; Pokines and Higgs 2015). However, the majority of empirical studies on postmortem activity by crustaceans have been conducted on non-human models (Anderson and Hobischak 2004; Anderson and Bell 2014, 2016). Crustaceans, including crawfish, feed on accessible areas, causing lesions to exposed body regions that are uncovered by clothes, such as the abdomen and open orifices, either natural (e.g. nose, mouth, and ears) or artificial (e.g. perimortem wound) (Haglund 1993; Hobischak and Anderson 2002; Anderson and Hobischak 2004; Duband et al. 2011). Species of crawfish are known for their opportunistic scavenging behaviour on marine faunal species and human cadavers (Anderson and Bell 2014, 2016). Crustaceans mostly live on the seafloor and are more likely to scavenge human remains lying there than those floating or drifting (Anderson and Hobischak 2004), as observed in the case of the 2009 Air France airplane crash. Studies focussing on the interaction between molluscs and decomposing remains showed that the impact is more superficial than that of arthropods because generally it is the skin that is affected. However, by nesting or simply attaching to the skin, some molluscs create patterns that may resemble perimortem trauma, such as 'stab wounds' from sea lice (Byard et al. 2002) or marks from bivalve molluscs, such as *Xylophaga dorsalis* (Introna et al. 2013; Pokines and Higgs 2015). There is, therefore, a high degree of complexity that needs to be considered when seeking to correctly identify the cause of these patterns. For example, it is still unclear whether the postmortem modifications *Xylophaga dorsalis*, a wood eater, leaves on human remains actually correspond to its feeding attachment in which it anchors to the human skin, or whether the mollusc actually feeds on the human remains like it does on wood.

Larger Fauna

The impact that larger fauna have in the decomposition process of human remains is more described in the published literature than that of small fauna. Various species of marine vertebrates, including fish, rays, and sharks, are known to interact with humans, ante-/perimortem and postmortem (Taylor et al. 2002). Among these, the impact of shark action on human remains has been particularly studied.

A number of shark attacks are reported worldwide each year (Stock et al. 2017) and are of great interest specifically in the areas of public security and health. For instance, 126 shark attacks were reported in the year 2016 only, but not all of them were fatal or involved any injury at all (Shark Research Institute 2017). This highlights the importance of collaboration between medical disciplines and forensic archaeology and anthropology. The published literature distinguishes between two types of shark bitemarks: deep, large, V-shaped marks with irregular edges, some of which show puncture wounds (tooth notches) around the edges (Fig. 9.9a, c), and the small incised wounds arranged in a semi-circle (Fig. 9.9b) (Ritter and Levine 2005; Motta and Wilga 2001; Huber and Motta 2004; Byard et al. 2002, 2006). The first are generally considered as complete bitemarks, whereas the latter are often interpreted as incomplete or unsuccessful bitemarks (Ritter and Levine 2005; Motta and Wilga 2001; Huber and Motta 2004; Byard et al. 2002, 2006). Shark-inflicted incised wounds may be misinterpreted for perimortem human-inflicted injuries, such as stab wounds (Byard et al. 2002). Lesion depth determines whether the trauma was lethal, which is essential in the diagnosis of the cause and manner of death. When the lesions are deep, the presence of tooth notches can corroborate the diagnosis of shark bites, thereby limiting the risk of uncertainty in postmortem conclusions (Byard et al. 2002). The impact of the scavenging activity of large sharks has been much more extensively studied than that of smaller sharks (Jones et al. 1998). Despite this, small and even "dwarf" species of sharks are thought to be important taphonomic factors (Ribéreau-Gayon et al. 2017, 2018).

Several reports in the literature mention bitemarks on humans from the so-called cookiecutter sharks (*Isistius* spp.) (Gunatilake 1992; Shirai and Nakaya 1992; Gadig and Gomes 2002; Makino et al. 2004; Papastamatiou and Wetherbee 2010; Honebrink et al. 2011; Hoyos-Padilla et al. 2013; Claes et al. 2014; Hayashi et al.

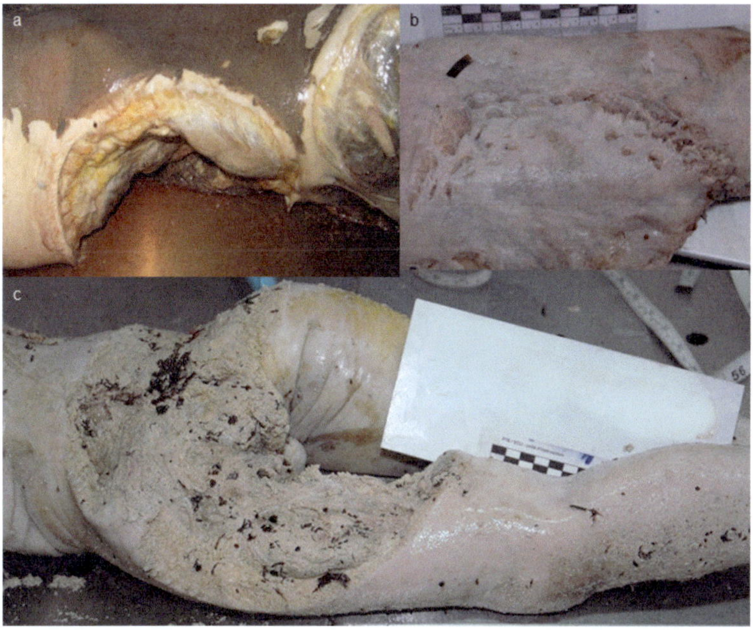

Fig. 9.9 Large shark bitemarks on the victims of the Yemenia airplane crash (29th June 2009). IRCGN (**a** and **c**) Deep, large, V-shaped bitemarks with tooth notches on the edges. (**b**) Superficial, small, incised semi-circular bitemark

2015). *Isistius* spp. have a unique feeding behaviour known as kleptoparasitism; the shark removes plugs of flesh from its prey using its suctorial lips, sharp teeth, and very strong throat muscle. This leaves scooped-like circular bitemarks, as shown in Fig. 9.10a, b. However, the majority of studies available are exclusively based on isolated cases. An empirical study on a large cohort of human victims (n = 113) identified 560 cookiecutter bitemarks, the majority of which (29.1%) was located on the anterior thorax, as shown in Fig. 9.10c (Ribéreau-Gayon 2014; Ribéreau-Gayon et al. 2017). This study highlighted that cookiecutter sharks have an important impact on the degradation of human remains immersed in marine environments and, thus, are taphonomic agents with forensic significance in oceanic environments, something that may have been previously underestimated in forensic investigations that involved marine environments.

Postmortem predation by other species of small sharks is suspected, but has not been established yet (Dr Bernard Seret, pers. comm.). For instance, potential interactions between *Dalatias licha* (the kitefin shark) and human remains are poorly understood. It is believed that *Dalatias licha*, which resembles a large cookiecutter shark, but is approximately three times longer (approximately 180 cm) (Compagno et al. 2005), also feeds via kleptoparasitism, leaving circular bitemarks that are larger than those of *Isistius* spp.: about 8 or 10 cm long compared with about 4 or 5 cm for *Isistius* spp. (Dr Bernard Seret, pers. comm.). These preliminary hypotheses need to be validated by empirical studies, and the potential forensic importance of species of "dwarf sharks", such as *Euprotomicrus bispinatus* ("pygmy shark") and *Squaliolus* spp. (*S. aliae* and *S. laticaudus*), also needs to be addressed.

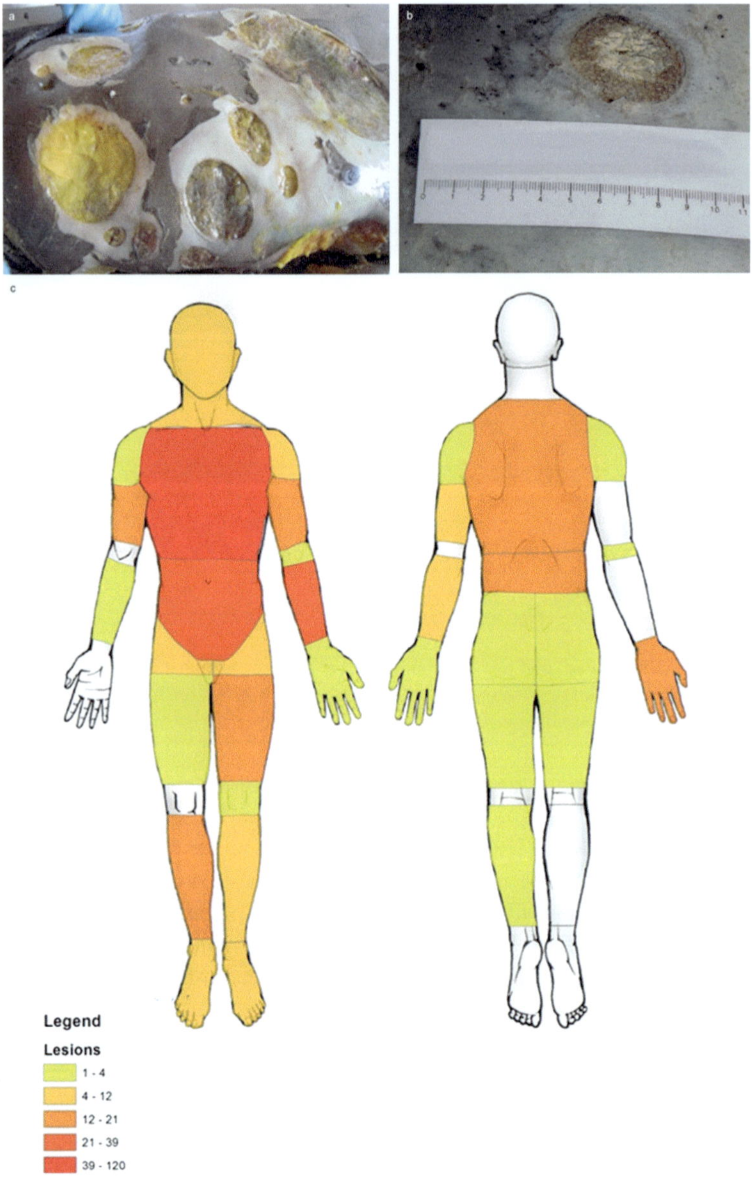

Fig. 9.10 Macroscopic appearances of *Isistius* spp. bitemarks on the victims of the Yemenia (**a**) and Air France (**b**) airplane crashes (29th and 1st June 2009). IRCGN. Frequencies and distribution patterns of the bitemarks identified in the case of the Yemenia airplane crash (**c**) (Ribéreau-Gayon 2014)

The Presence of Garments and Body Coverings

The presence of garments greatly appears to preserve immersed human remains, including in preserving the soft tissues, and contributes to maintain the anatomical connexions, as well as preventing the access of scavenging fauna to the remains (Introna et al. 2013; Ribéreau-Gayon et al. 2017).

Case Study

In the case of the Air France airplane crash, it was found that tight clothing, such as underwear, stockings, tights, and shoes, enabled particularly long-lasting preservation of the tissues, even after 24 months of immersion, as shown in Fig. 9.11. This is all the more noticeable in comparison to body areas that were exposed, such as the lower limbs that skeletonised faster, as illustrated in Fig. 9.11b. Body coverings, including clothing, blankets, and carpets, can also lead to the formation of adipocere (Notter and Stuart 2012). The conditions that lead covered body parts to be preserved or to become adipocere are not fully understood to date, and the case of the Air France airplane crash provides a very interesting example when both adipocere and non-adipocere covered remains were observed in similar environmental conditions and PMSIs.

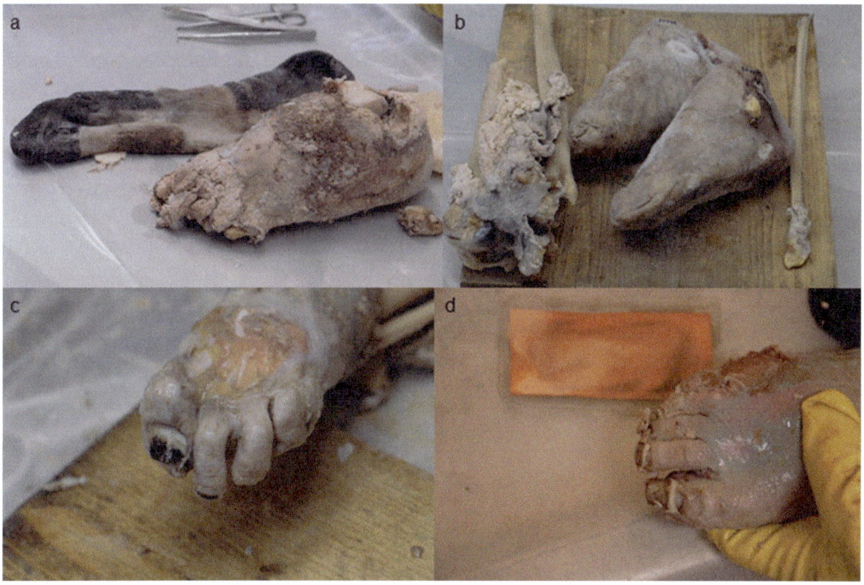

Fig. 9.11 Examples of preservation of soft tissues enabled by garments, Air France crash (1st June 2009). IRCGN (**a**) Coverage by socks and stockings preserved the feet of the victims after 2 years of immersion in the Atlantic Ocean. (**b**) Note the remarkable difference in the level of preservation between the feet and the rest of the lower limbs of the individual. (**c** and **d**) In some cases, the toenails were even preserved

9.4 Perspectives on Forensic Investigations in Marine Environments

The majority of the current evidence base of forensic marine taphonomy has been established from anecdotal reports and casework, rather than from empirical research (Stuart and Ueland 2017a). However, empirical research can provide essential knowledge of the complex interactions between human remains and their depositional environment, which enables the increase of the chance of the recovery of the remains and the reconstruction of the sequence of events from time of death to time of discovery and interpretation of the remains. More specifically, empirical research provides the means to understand in which forms, quantities, and locations certain types of evidence, such as human remains, can be found and to evaluate their evidentiary value. This approach increases the chances of positive identification of the deceased.

This chapter illustrates how large datasets have the potential to enable the development of a specific evidence base that can allow us to make evidence-based inferences in subsequent forensic reconstructions (Sects. 9.2 and 9.3). Indeed, large-scale retrospective studies on datasets with known data, including from airplane crashes and boating accidents (Kahana et al. 1999; Campobasso et al. 2006; Introna et al. 2013), offer valuable opportunities to study large number of victims that span various taphonomic stages (Heaton et al. 2010; Dabbs et al. 2017). Additionally, datasets that represent different age ranges, heights, body masses (BMI), and biogeographic origins (e.g. highly pigmented dark complexions and poorly pigmented light complexions) are all the more valuable as they reflect decomposition and taphonomic patterns in marine environments more accurately than a small selection of human remains would. For example, a potential challenge of postmortem investigation that has been identified in the literature is whether or not taphonomic patterns, such as hypostasis (or lividities), are visible on the most pigmented dark skins (Lawler 1992). Such datasets have the potential to provide significant contributions to the study of marine taphonomy to build upon the majority of the published literature that currently focusses on isolated casework or experimental studies addressing non-human materials.

Decomposition in marine contexts is complex given the multiple variables within these environments (Kahana et al. 1999; Morgan et al. 2006; Introna et al. 2013; Schuliar et al. 2014; Brooks and MacDonald 2014; Ribéreau-Gayon et al. 2017; Ellingham et al. 2017). As such, the field of marine taphonomy is inherently interdisciplinary and thus requires knowledge and expertise from experts from various fields such as underwater archaeology, forensic anthropology, forensic pathology, forensic science, marine biology, geology, etc., as illustrated in Fig. 9.12. Collaboration between forensic science researchers and practitioners is essential to ensure a sufficient awareness of the challenges that aquatic environments entail and that research questions and methods are crafted in a manner that is applicable and implementable in the "real-world" context (Morgan 2017b; Howes 2015).

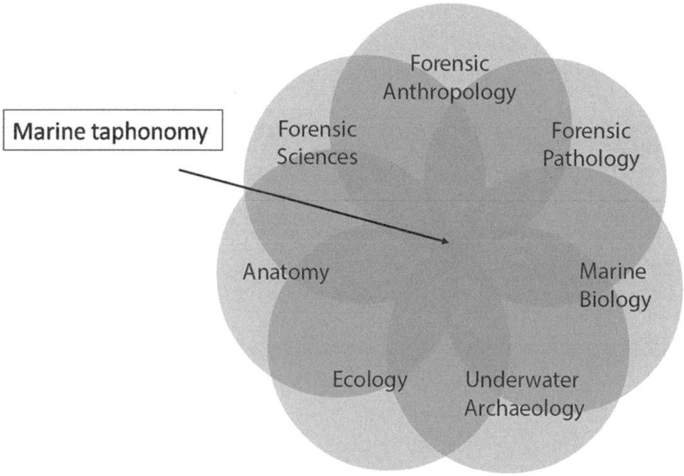

Fig. 9.12 Marine taphonomy at the crossroads between various connected disciplines

Collaborative approaches can help to ensure clear and helpful communication between different institutions and actors which in turn contributes to the effective investigation and reconstruction process of the series of events from time of death to time of recovery.

9.5 Conclusion

This chapter has outlined a number of the challenges encountered in forensic archaeology and anthropology investigations in marine environments, from detecting the remains to interpreting their significance. The complexity of the taphonomic process is driven by a number of factors, both exogenous and endogenous, which can leave various taphonomic patterns on human remains at both macro and micro scales. The ability to accurately analyse these patterns and reconstruct the taphonomic sequence varies depending on the extent of the evidence base available. The key gaps in our understanding of these taphonomic processes concern the types of taphonomic factors that are acting in marine environments and their precise impact on the decomposition process of human remains. More precisely, the range of macroscopic expressions of the taphonomic patterns as well as their onsets and durations remain unclear which currently introduces uncertainty in the reconstruction of the depositional environment. Thus, interdisciplinary empirical research that incorporates both practitioners and researchers is essential to better understand the issues pertaining to decomposition in marine environments and to develop robust approaches at each stage of the investigation process to enable more accurate forensic reconstructions in practice. The importance of developing generalisable theories that can be applied in a context-sensitive manner is clear, and developments in our ability to reconstruct the taphonomic process in marine environments have significance not only for aquatic forensic investigations and the establishment of robust PMSI indicators but also for forensic reconstruction approaches in other environments.

Acknowledgements The authors are grateful to the Forensic and Criminal Intelligence Agency of the French Gendarmerie (IRCGN) for providing access to the materials on the accidents presented here.

The authors respectfully acknowledge the families and relatives of the victims of the accidents.

Chapter 10
Interrogating the Ground: An Archaeologist at a Crime Scene

Tomasz Borkowski and Maciej Trzciński

Abstract *Nescire autem quid ante natus sis acciderit,* id est *semper esse puerum. Quid enim est aetas homini, nisi ea memoria rerum veterum* cum *superiorum aetate contexitur?* (Cicero 46 BC).

This paper deals with essential issues associated with Polish forensic archaeology, ones which are closely linked with recent history. Hence, the vast majority of actions undertaken concern the search for victims of crimes against peace, humanity and war crimes, with a particular focus on crimes committed by the communist regime. Polish forensic archaeology, in terms of the tasks which it faces, differs from the Western European countries, where archaeological methods are used primarily to search for murder victims and the concealed proceeds of crime, as well as to perform visual inspections of mass disaster scenes.

Keywords Forensic archaeology · Communist regime · Poland

10.1 Introduction

The authors of this article have been involved for several years in work associated with the search and exhumation of victims of the communist regime. As this work is carried out in diverse environments, i.e. in high-density urban areas, active cemeteries and historical buildings as well as in undeveloped areas, very diverse sources of information are encountered in the course of these investigations concerning the location of the concealed human remains. The characteristics of the given areas ensure that the work methodology applied is unique to each site. Formal and legal aspects of a forensic archaeologist's co-operation with the police, prosecutor's office and heritage protection office are also discussed in this paper.

T. Borkowski (✉)
Institute of National Remembrance, Warsaw, Poland
e-mail: tomasz.borkowski@ipn.gov.pl

M. Trzciński
Faculty of Law, Administration and Economics, Department of Forensic Sciences, University of Wrocław, Wrocław, Poland

P. M. Barone, W. J. M. Groen (eds.), *Multidisciplinary Approaches to Forensic Archaeology*, Soil Forensics, https://doi.org/10.1007/978-3-319-94397-8_10

The problems associated with Polish forensic archaeology are illustrated by the extraordinary case of a search for the murder site and the location of the concealed remains of the soldiers of a Polish underground unit, who were killed in September 1946.

10.2 Forensic Archaeology

The archaeology of contemporaneity, sometimes called the archaeology of the recent past, has been engaging in research areas, which have not previously been investigated by Polish archaeologists, in an increasingly more dynamic and courageous manner in recent years.[1] The reason for this is that there are new specialisations that have arisen as a response to new directions in the research activity of archaeologists. These areas of research are well established in Western European countries and the USA, such as battlefield archaeology[2] or the increasingly frequent research of late modern and contemporary times.

Dramatic events in the recent history of Poland resulted in the need for archaeology to face the issue of war crimes[3] and, recently, with increasingly frequent searches for victims of the communist regime[4] and their subsequent identification.

Forensic archaeology maintains its distinctiveness due to the scope of application. It is closely linked to investigations, which are aided by archaeological methods and techniques. Forensic archaeology differs from archaeology as it has been traditionally understood in that archaeology and archaeologists become a 'tool' in the hands of a judicial body, usually a prosecutor, or the police. Thus, forensic archaeologists do not conduct traditional archaeological research, but rather participate in acts performed in relation to an ongoing investigation. It is worth noting here that, under the existing legal regulations, archaeologists may be involved in an ongoing investigation as expert witnesses (Article 193(1) of the Code of Criminal Procedure)[5] or as specialists asked , e.g. to perform certain activities in connection with the visual inspection of a site (Article 205(1) of the CPC). Thus, the role of forensic archaeologist in the proceedings is provided for by law. It should be stressed here that the effectiveness of the forensic archaeologist in an ongoing investigation depends on the communication between the prosecutor (or the police) and the forensic archaeologist.

---

[1]This issue was first noted at the First Congress of Polish archaeology, held in Warsaw on 19 September–21 September 2013. In the aftermath of the conference, *Archeologia współczesności* was published (Warszawa 2016).

[2]See *Schlachtfeldarchäologie. Battlefield Archaeology*, Halle 2009; T. Lynch, J. Cooksey, *Battlefield Archaeology*, Chalford 2007.

[3]See, e.g. A. Kola, *Archeologia zbrodni. Oficerowie polscy na cmentarzu ofiar NKWD w Charkowie*, Toruń 2005.

[4]*Archeologia totalitaryzmu. Ślady represji 1939–1956*, Łódź 2015.

[5]Act of 6 June 1997 – *Code of Criminal Procedure*, Journal of Laws of 1997 No. 89, item 555, as amended.

10.3 Research Methodology

Despite the above comment that forensic archaeologists rather participate in activities as requested by the prosecutor or the police than conduct archaeological research, forensic archaeology itself should be considered a scientific discipline which might not be autonomous, but has surely established its distinctiveness.

It should be noted that the legal definition of archaeological research has been included in Article 3(11) of the *Act of 23 July 2003 on the protection of monuments and the guardianship of monuments*, where we can read that archaeological research are 'activities aimed at discovering, identifying, documenting and securing an archaeological monument'.

Taking into account the fact that forensic archaeologists participate in the activities initiated by law enforcement authorities, they should familiarise themselves with the necessary basics of the penal procedure, as well as the legal regulations concerning the performance of visual inspections because it is these regulations that directly determine the phases and standards of conduct with which the police must comply.[6]

The activity of forensic archaeology focuses on several fundamental topics. These include primarily (1) searching for victims of crimes against peace, humanity and war crimes, (2) exploration of the burial places of such victims, (3) searching and exploration of burial places of victims of contemporary crimes, (4) performance of expert visual inspections together with documentation on the scenes of transport disasters and (5) archaeologists' involvement in matters concerning crimes against cultural heritage.

In the last few years in Poland, forensic archaeology has been an important part of the searches and exhumations carried out by the Institute of National Remembrance in connection with investigation of crimes of the communist regime.

Archaeologists can be summoned to the crime scene when there is a need for specialists dealing with features which are beneath the topsoil. In such a situation, there is little difference between what they do on archaeological sites during 'standard excavations' and how they can contribute to a crime scene investigation. This has to do primarily with the specific way of dealing with features as well as the layers and objects found in them. Every archaeologist is a detective, and an excavation is 'an interrogation of the ground'. Archaeologists 'question' the earth because it is a witness of the events of the past (in many cases the only witness, which is an objective one – the ground does not lie). Whether the recording of features is performed with the use of traditional methods, a measuring tape, a level, a pencil and a graph paper, or with the aid of the most modern electronic devices, it is merely a matter of comfort, precision and speed. What actually matters the most is the archaeological thinking about the past and the knowledge and experience in this

[6] *Guideline No. 1 of the Chief of the Police of 23 July 2015 on the performance of certain intelligence collecting activities by police officers*, Dz. Urz. KGP of 2015, item 59, especially Chapter 8, *Oględziny miejsca, osoby, rzeczy* (Visual inspection of places, persons and items).

field. And whether an archaeologist is trying to understand the motives of a 'perpetrator' of a wattle fence from the Middle Ages or a perpetrator of a contemporary crime makes very little difference because both events occurred in the past and traces thereof remain in the ground, and it is possible to read and interpret them with the use of the proper methods. Naturally, contemporary crimes evoke stronger emotions, which are likely where the most important difference between 'normal' archaeology and forensic archaeology lies. Archaeologists dealing with the contemporary times are exposed to greater pressure. Moreover, co-operation with a prosecutor's office or the police can be stressful (haste, misunderstandings resulting from application of different procedures, ignorance of the law on the part of the archaeologist and ignorance of archaeology on the part of the other party).

In our opinion, the methods of treating the ground used in forensic archaeology are virtually the same as in the case of classic archaeology. One example that can be used to illustrate this is the hut of a former German military airfield in Alt Grottkau, currently Stary Grodków in Opole Voivodeship, Poland.

10.4 The Airfield in Stary Grodków/Alt Grottkau Case

The airfield in Stary Grodków/Alt Grottkau today is a largely forested area, although its remains are still perceptible. There are old roads, kerbs, water tanks, concrete structures, hut foundations and a lot of artefacts. As a result of fighting in 1945, the Alt Grottkau airfield was heavily damaged, but some structures survived. It was not used by the Soviet or Polish armies, but just abandoned.

In September 1946, more than a year after the end of World War II, according to written sources, documents, witnesses, rumours, etc., several dozen soldiers of the Polish underground, who were fighting against the new communist regime, were murdered there by Polish Security Service officers. They were supposed to have been blown up in one of the huts by anti-tank mines that had been planted under the floor (there was space between the ground and the floor of about 40 cm).

The soldiers belonged to the National Armed Forces; their commander was Captain Henryk Flame, pseudonym Bartek. The unit in question was a very well trained and armed. There were more than 300 soldiers in the unit. They operated in their native highland area (the Carpathian Mountains), having nearly 100% support of the local community. The Security Service prepared 'Operation Avalanche' in order to lure them out from their locality, where they were practically untouchable. An agent provocateur convinced some of the unit's members that they could be transferred safely to the West. The commander, having his doubts, let the soldiers decide on their own. Some of them left, some stayed. One of transports (they travelled on trucks, armed) arrived in Alt Grottkau (a distance of about 200 km). They stopped for the night in the abandoned airfield. When they went to sleep, the structure was blown up with them inside.

The search in Stary Grodków began in 2015. It was ran by the Institute of National Remembrance in co-operation with the Museum of Archaeology in Wrocław. In the initial stage, geophysics (GPR) and non-invasive surface prospection were carried out. The geophysics research did not provide any information. The surface prospection was followed by a metal detector search, which lead to the identification of several clusters of artefacts, the most interesting of which seemed to be a concentration located in the northwest corner of a destroyed airfield hut (Fig. 10.3). Amongst the German and Soviet World War II objects recovered were several British battledress buckles, which were an anomaly, indirectly pointing to Polish underground soldiers, who used British equipment.

The hut was almost completely destroyed, and only the foundations (Fig. 10.1) and various elements of equipment scattered in the ground in the immediate vicinity, including a significant number of melted shards of window panes (Fig. 10.2), survived until the time of the excavations which were carried out in 2016. Based on the preserved fragments, it was possible to determine the measurements of the hut, as well as its construction and function (Fig. 10.3). The remains of the hut were subject to such archaeological research as is the case for the remains of a large archaeological feature (e.g. a Neolithic longhouse or a megalithic structure). The

Fig. 10.1 Stary Grodków/Alt Grottkau. Former military airfield. Hut's foundations. (Photo by T. Borkowski)

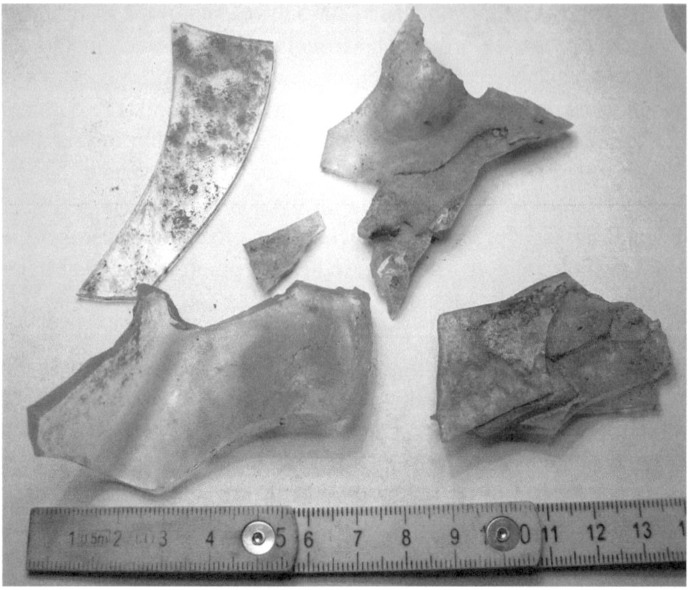

Fig. 10.2 Stary Grodków/Alt Grottkau. Former military airfield. Fused fragments of window panes discovered in the vicinity of the hut. (Photo by T. Borkowski)

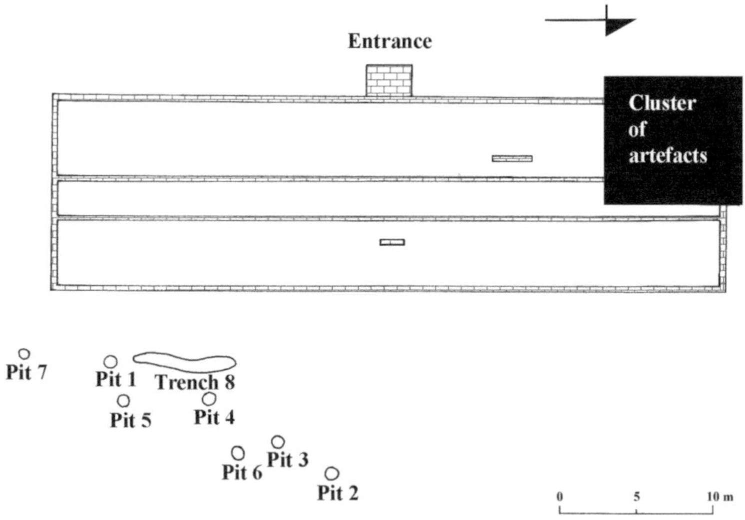

Fig. 10.3 Stary Grodków/Alt Grottkau. Former military airfield. Plan of the hut. (Drawing by T. Borkowski)

fused window panes' fragments were evidence of high temperatures at the time of the destruction of the facility (fire). The brick foundations were not particularly solid, which suggested that the building was a light, temporary structure. Naturally, such huts remain to this day; it is also possible to find photographic documentation showing such buildings from the time when the airfield in Stary Grodków was operational. However, if we did not have such information, we would have to base the results on archaeological work (which happens frequently).

In Spring 2016, the excavation work started, initially in the area of the artefact cluster. Again, several British buckles were found, but these were still only clues. Soon they were followed by Polish pre-war coins, a boy scout cross, military emblems, cartridge cases which exploded in the fire (not in the barrel) and finally by a metal gorget typical of the soldiers of the unit in question.

Knowing that we were close, we started open-air excavations (Fig. 10.1). On the opposite side of the hut (next to its southeast corner), seven pits and one trench containing partially disarticulated human remains were found. The pits were circular in plan and conical in section. They usually contained some rubbish in the bottom (iron rods, broken glass, etc.). Taking into account that only seven of the pits contained human remains and we found about 40 of them – all carefully exposed by a mechanical digger – we assumed that perpetrators had used existing pits, very likely mortar bombardment craters (Figs. 10.4 and 10.5).

Fig. 10.4 Stary Grodków/Alt Grottkau. Former military airfield. Probable mortar bombardment craters during excavations. (Photo by T. Borkowski)

Fig. 10.5 Stary Grodków/Alt Grottkau. Former military airfield. Part of World War II ammunition (including mortar shells) found during the excavations. (Photo by T. Borkowski)

The largest number of human remains was found in trench No. 8 (Figs. 10.6 and 10.7). It was discovered about 0.40 m below the current ground level. In the upper part, there was burnt material, charcoal and various iron objects – most likely the remains of the burnt-down hut which might have concealed the burial place. The trench was 6.32 m long, the maximum width being 1.21 m and the maximum depth being 1.10 m. The human remains discovered in the trench had traces of various injuries, including gunshot ones (Fig. 10.8). The items found with the skeletons (e.g. another gorget) indicated that they had been Polish underground soldiers from the unit of Captain Henryk Flame. Due to the use of archaeological methods, a fragment of a newspaper with a handwritten note near one victim was found, which might help in their identification. In trench No. 8, apart from a syringe, an undetonated grenade, a brass cigarette case, fragments of rosaries, a signet ring with initials, etc., 12 bullets were also found. The presence of bullets indicates that after the explosion these soldiers were finished off with bullets, probably because they were armed and they were considered very dangerous.

Fig. 10.6 Stary Grodków/Alt Grottkau. Former military airfield. Trench No. 8. (Photo by T. Borkowski)

Fig. 10.7 Stary Grodków/Alt Grottkau. Former military airfield. Trench No. 8. (Photo by T. Borkowski)

Fig. 10.8 Stary Grodków/ Alt Grottkau. Former military airfield. Human bone with a gunshot injury discovered in trench No. 8. (Photo by T. Borkowski)

10.5 Conclusions

Based on very weak evidence, such as testimonies of witnesses describing events from 70 years ago, rumours and testimonies of the perpetrators themselves (undoubtedly biased and often simply untrue), an approximate search location was selected. The use of the archaeological method enabled the determination of facts, and the ground subject to 'interrogation' constituted an ideal, objective witness.

Chapter 11
Forensic Archaeology in Denmark: A Work in Progress

Lars Krants

Abstract There has been formalised collaboration between the Danish National Police and Moesgaard Museum since 2010. While this relationship was initially characterised by happenstance, through personal contacts and close dialogue, it has since undergone progressive development. The cooperation relates to actual criminal investigations but, to an equal degree, also encompasses knowledge sharing through excavation courses and lectures, so that both police and archaeologists are professionally better equipped and qualified in relation to these cases. This close working arrangement is under continual refinement, both professionally and methodologically, and has since 2015 acquired an international aspect in the form of cooperation under the auspices of the European Network of Forensic Science Institutes (ENFSI). A brief account is given here of the Danish museum and police organisations, forensic archaeology and crime statistics in Denmark and the practical nature of the cooperation between archaeologists and the police. The article concludes with a presentation of a number of cases that shed light on various tangible aspects of the collaboration and an assessment of what the future holds.

Keywords Forensic archaeology · Historical overview · Crime scene investigation · Viewshed analysis

11.1 Introduction

Forensic archaeology is a recent arrival in Denmark, and its application was initially characterised by personal contacts and happenstance. Moreover, as there is no official forensic archaeological training available in Denmark, and relevant cases are relatively few and far between, it is crucial to maintenance of standards and method

L. Krants (✉)
Moesgaard Museum, Højbjerg, Denmark
e-mail: lkl@moesgaardmuseum.dk

P. M. Barone, W. J. M. Groen (eds.), *Multidisciplinary Approaches to Forensic Archaeology*, Soil Forensics, https://doi.org/10.1007/978-3-319-94397-8_11

development that it is the same people who undertake the individual investigations. Consequently, a formal collaboration has been established between the Danish National Police, Moesgaard Museum and the Department of Forensic Medicine at the University of Copenhagen. In order to facilitate an understanding of the interaction between archaeologists and the police, the two organisations are outlined below. For information on the Department of Forensic Medicine see Jørkov and Lynnerup (2015).

11.2 Archaeology and the Danish Museum Service

In Denmark, archaeological investigations are undertaken by 27 state-authorised cultural-historical museums, each of which is responsible for a specific geographical area. Collectively, these areas of archaeological responsibility cover the entire country, and the museums have the exclusive right to undertake archaeological investigations. The investigations are carried out by professional archaeologists trained at either Copenhagen or Aarhus University. Most archaeological activities take the form of so-called rescue excavations, which are undertaken in connection with various kinds of development or construction work and are subject to directives in the Museums Act. As a point of departure, it is always the developer in question who must finance the archaeological investigation. Because the museums have a monopoly-like status in this respect, and as the developers are not free to choose who they will work with, the entire area is subject to continual monitoring by a special agency under the Ministry of Culture, which must approve all investigations and ensure that professional standards are upheld. In parallel with this rescue archaeology, most museums also engage in various kinds of archaeological investigations and research excavations which are usually funded by private or state foundations.

Moesgaard Museum, where the author is employed, is one of the aforementioned 27 museums with archaeological responsibility. In addition to archaeology on land, the museum also has responsibility for the marine-archaeological heritage (jointly with four other museums) and for monitoring scheduled ancient monuments (jointly with nine other museums). The museum has close and formal collaboration with Department of Archaeology and Heritage Studies at Aarhus University, with whom it shares premises and office facilities. Even though it receives public funding from both the state and the local municipality, the museum is an independent self-governing institution and has its own governing body. This confers a greater degree of flexibility relative to museums that are municipally or state-owned, and this situation has proved to be particularly advantageous in the formative phase of the collaboration between the police and the museum.

11.3 Organisation of the Police

The police in Denmark, the Faroe Islands and Greenland are all part of one police force under the Danish Ministry of Justice, with the incumbent Justice Minister as their ultimate authority. The day-to-day running of the force is undertaken by the National Police Commissioner, together with a number of chief constables. With the implementation of a major police reform in 2006, 54 police districts were reduced to 14, 12 in Denmark, 1 on the Faroe Islands and 1 in Greenland, but the Danish National Police constitute the supreme Danish police authority. The Danish National Police are responsible for the formulation of strategies and for defining the frameworks and direction of the police's activities and for the undertaking of work on a national basis. In day-to-day activities, however, it is primarily the individual police districts that set the agenda, and there can therefore be a relatively large degree of autonomy between them.

In the event of more serious crimes, the individual police districts are though obliged to contact the Danish National Police. The latter have a dedicated support function in relation to criminal investigations in the police districts and are responsible for ensuring that national strategies are followed. Similarly, it is the Danish National Police who supply the necessary specialist competences and possess the methods, tools and techniques that are necessary for the investigative work.

Forensic archaeology and archaeologists are part of the Danish National Police's "toolbox", and forensic archaeology and anthropology can be included in the detection work to the extent that the Danish National Police consider relevant. As each police district has its own budget, which they obviously take care to manage closely, problems of authority may arise if the Danish National Police recommend or require the use of forensic archaeological support in a particular case, because it is not the Danish National Police who must finance this support, but the actual police district involved.

11.4 Forensic Archaeology

It is not possible to train specifically as a forensic archaeologist in Denmark as there is no dedicated course. Neither is there an official job title of forensic archaeologist, either in the museum world or within the police force or the judicial system. It is up to the police, or rather the Crown Prosecution Service, to evaluate whether information and indications (circumstantial evidence) resulting from a forensic archaeological investigation are to be included as evidence in potential legal proceedings. But as there are no rules, practices or legislation defining who can be employed as an expert witness, for example, a self-taught forensic archaeologist with no official accreditation like the author, it is exclusively up to the judge in each individual case to assess whether an expert can be used as a witness.

11.5 Crime Statistics

At present, it is unfortunately not possible to obtain a clear and manageable overview of criminal cases in Denmark, including those involving homicide. Denmark's central statistics office (Statistics Denmark) extracts information via the Police Record System (POLSAS), but the figures are subject to a degree of uncertainty, for several reasons: For some periods, including 2001–2009, there are simply errors in the data extracted from POLSAS,[1] but the police's way of recording cases involving homicide also has an influence. For example, a corpse will only be recorded as a homicide case if lesions (i.e. injuries) can be demonstrated that can unequivocally be linked to the cause of death. This applies even if the body has been buried! In other words, the official statistics contain a number of dark figures, and the actual number of homicide cases is probably slightly greater than that indicated by the official statistics. Despite these caveats and various statistical uncertainties, the figures display a remarkable consistency: They show that, over the last quarter of a century, there has been very precisely c. 1 homicide per 100,000 inhabitants per year (the Danish population is currently c. 5.7 million).

In addition to the information on criminal offences extracted by Statistics Denmark, the Research Division of the Ministry of Justice produces reports and statistics and undertakes research within a broad area of the ministry's activities, including a forensic approach. All their reports are freely accessible on the Ministry of Justice's web page, but unfortunately in most cases, they are written in Danish with no English summary.[2] There is currently no official report on homicide cases where the corpse was completely or partially buried. However, the author expects this situation to be rectified in the near future because Asser Hedegård Thomsen, specialist in forensic medicine at the Department of Forensic Medicine, Aarhus University, has, as part of his PhD thesis, examined all homicide cases from 1992 to 2016.[3] From his preliminary research, it is evident that in approximately 1.7% of all recorded cases of homicide, the victim was buried, giving an average of one case per year. Bearing in mind the aforementioned statistical uncertainty, and that a number of people are reported missing each year, some of whom may have been killed and buried but have not yet been discovered, it is not unreasonable to conclude that there are, on average, one or two cases each year where the victim has been buried. The distribution of this special kind of homicide case does not, however, appear to be even over time, as no clandestine graves have been discovered during the last 7 years. But this does not mean that there have been no cases where the victim was buried. Several individuals have been reported missing during this period, and some of these must, based on information and circumstantial evidence, be presumed to have been killed and possibly buried but have yet to be discovered. For examples of these, see below.

[1] https://www.dst.dk/da/Statistik/dokumentation/statistikdokumentation/ofre-for-anmeldte-forbrydelser/sammenlignelighed

[2] http://www.justitsministeriet.dk/arbejdsomraader/forskning/rapporter-vedr-forskningspuljen

[3] Asser Hedegård Thomsen is thanked for the following information.

11.6 Collaboration Between Police and Archaeologists

There has been formal collaboration between the Danish National Police and Moesgaard Museum since 2010. Prior to this, however, there were several cases where police and museums worked together. For example, Nordjyllands Historiske Museum in Aalborg and Museum Lolland Falster in Nykøbing Falster have both occasionally aided police investigations. Cooperation between the police and Moesgaard Museum arose, as such things often do, almost by chance. A colleague, who is head of the Department of Environmental Archaeology and Conservation at Moesgaard Museum, has an acquaintance in the police who he invited, together with his colleagues, for a guided tour. His clandestine hope was that the police would perhaps be interested in making use of some of the department's classic archaeological skills in connection with their investigative work, for example, pollen analysis or plant macrofossil analysis. To flesh out the programme, the author was asked to give a lecture about forensic archaeology and the positive contributions that archaeologists could perhaps make to the work of the police. The lecture, naturally, dealt primarily with how archaeology can help in relation to buried corpses, but it also touched upon survey work, use of drones and the application of geophysics. Noattempt was made to hide the fact that author's experience of police work was strictly theoretical and mostly based on reading a book by Hunter and Cox (2005) that was in the museum's library. In conclusion, it was mentioned that the author would of course be very willing to assist with future cases and that he would encourage the police to find a collaborative partner in whom they had confidence and then make consistent use of them in future cases, thereby ensuring the professional upgrading of competences and development of the field.

The lecture must, in some way or other, have made an impact, because shortly afterwards Moesgaard Museum was invited to become involved in the first police cases. In the beginning, the initiative came from the local police districts, but very quickly the collaboration began to go via the Danish National Police's official expert in burial cases,[4] and subsequently it became significantly more targeted and focused. One of the first actions was to establish contact and begin collaboration with the Danish National Police Dog Training Centre[5] and jointly setting up a series of courses in excavation techniques for forensic officers. The aim of these courses was not that the police should take over the work of the archaeologists but that they should become familiar with the basic principles behind archaeological practices so that they are able, on a qualified basis, to make demands and requests in connection with individual investigations. The first courses began in 2014, and by the end of 2017, all the forensic officers in Denmark had participated.

[4] Police inspector Henrik Hougaard Jensen, Danish National Forensic Service (NKC).

[5] Detective inspector Jette Hardrup from NKC has been responsible for contact to the Danish National Police Dog Training Centre, initially via Ove Danielsen and since 2014 via police inspector Steen Stausholm.

The courses were, from the outset, made as realistic as possible. Various objects, for example, suitcases containing clothes, blood and decomposition fluids from corpses, are buried well in advance of the start of the course. The special cadaver dogs, systematically used for this type of case in Denmark, then get to work. The dogs react to corpse odour and to burials in general. Indications can therefore come in response to other kinds of burials, for example, weapons, or, as has happened on one occasion, the dogs may react to an unused grave or a grave that for some reason or other has not been completely dug. In most cases the dog picks up the scent along the cut edges of the grave, but if it is sited close to a perimeter drain or a sewer or in particular types of soil, the scent can wander from the actual place of burial and several metres out to the side. With time, it becomes increasingly difficult for the dogs to locate the grave, and other more customary archaeological methods must be employed, for example, removal of the surface soil by an excavator. When the grave has been located, the area is cordoned off, and other dogs (i.e. tracker dogs) search the immediate surroundings for relevant objects that can be significant for resolution of the case. It is first when possible objects have been found and secured that an access route is established to the grave and archaeologists and forensic scientists/officers are allowed on to the scene.

In recognition of the fact that these courses are intended for people who have very little or no prior experience of archaeological methods, very simple excavation-related principles are employed. There is great emphasis on the fact that the work undertaken should be easily comprehensible visually and that the resulting evidence must ultimately convince a judge and possibly a jury who have perhaps never seen an archaeological excavation. The approach employed involves special use of surfaces, both vertical and horizontal, in order to demonstrate and visualise what is inside the grave and what is outside it. Similarly, we work with emptying the grave from one side rather than from the top. We are well aware of the fact that traces left by the digging implements on one of the sides of the grave may be disturbed as a consequence, but we believe that the resulting greater degree of visual clarity outweighs these disadvantages. That said, it is important to emphasise in the courses that it is crucial to maintain an open approach to excavation methods and practices: It is always the individual case and the police's investigation that should form the basis for the choice of excavation methods. If the situation or circumstances demand, methods should be changed or adjusted in the process, so they are more appropriate to the observed conditions. As archaeology is clearly a destructive process, it is important to employ working procedures which ensure that the excavation method can be changed underway, if conditions prove to deviate from those originally assumed. Consequently, procedures are applied whereby work begins by only excavating a part of the grave and does not continue until there is complete assurance that the conditions have been fully understood in the particular situation (Fig. 11.1). Part of the course naturally deals with recording practices, and there is training in 3D recording with the aid of a method known as Structure from Motion (SfM), which has the advantage that it requires no special equipment other than a camera. In time, however, there is little doubt the Danish police will also begin to make use of various forms of handheld 3D scanners.

Fig. 11.1 Forensic officers on an excavation course. First the grave is clearly identified in plan, after which it is excavated, beginning at the narrow end. Initially only one half of the grave is excavated. (Photo: Moesgaard Museum)

In parallel with the many courses and lectures, collaboration between the Danish National Police and Moesgaard Museum has become ever closer, and the forensic-archaeology group now encompasses five members from the Danish National Police and two from Moesgaard Museum, who meet on a regular basis.[6] This close collaboration is of crucial importance for the development of the field and has, since 2015, acquired an international perspective via participation in ENFSI. Moreover, both the police and the museum have participated in European Meetings on Forensic Archaeology (EMFA) and have taken part in, and arranged, excavation courses for European colleagues.

11.7 Presentation of Selected Cases

Over the last 7 years, there have been several cases where archaeologists from Moesgaard Museum have, in one way or another, aided the police: Brief accounts of several of these cases have been published previously. An account is given below

[6]In addition to the aforementioned Henrik Hougaard Jensen, Jette Hardrup and Steen Stausholm, the group comprises police inspectors Michael Fønsskov and Sidsel Nielsen, all from NKC, and museum curator Camilla Bjarnø from Moesgaard Museum.

of closed or current cases that give an impression of the nature and extent of this collaboration.

The case that has demanded the greatest resources to date is from northwest Zealand. It relates to a minor who was sexually abused by several of her family members and other involved individuals over a number of years. The investigations revealed conclusively that the girl had given birth to a child in secret. According to information from the abused, the child was born in her childhood home, killed by the family members and subsequently buried in the garden in an area between two trees. The witness also stated that the grave was about 1 × 1 m and of considerable depth, perhaps as much as 1 m, and that the baby was buried in a blue bag. The burial was said to have taken place 5–6 years prior to the beginning of the investigation in 2011. Before the actual excavation work, the area was scanned with ground-penetrating radar (GPR) by a private company. But these efforts gave very poor results, primarily as a consequence of the soil conditions but possibly also due to the use of outdated equipment. The indicated area was then investigated with a mechanical excavator, whereby very thin layers of soil were slowly and systematically removed (Fig. 11.2). An astonishing number of burials of household refuse were encountered, but most remarkable were the many animal burials: dogs, calves and even hens and chickens had been buried in the garden. The latter indicate that the soil conditions were of a character that even very small bones were preserved. There is consequently little doubt that even the bones of an infant would still be present in the soil. No grave was however encountered, even though the investigation area was massively extended. Ultimately, after three excavation campaigns, all accessible

Fig. 11.2 Forensic search for the grave of an infant. Eventually the entire plot was investigated. (Photo: Moesgaard Museum)

areas on the plot had been archaeologically investigated. Consequently, the conclusion had to be reached that, at the time of the investigations, there was no grave. Whether there had previously been a grave is unknown. It is possible that the infant's body had been moved on a later occasion or that the grave had been destroyed by animals (foxes?). Two men were remanded in custody, but due to the lack of evidence, the charges were subsequently dropped.

Another case relates to a young Syrian divorcee and mother of two who was reported missing in 2015. Various circumstantial evidence gave reason to suspect that her ex-husband, also from Syria, could have had something to do with her disappearance. Furthermore, it was presumed that she had been killed and perhaps buried: Newly acquired digging tools were found in the accused's possession which he had great difficulty in explaining. But then he was generally very incommunicative with respect to the police. Based on circumstantial evidence, various woods were searched with dogs. In 2016, the dogs reacted to a small grave-like pit in a small wood located south of Vordingborg, after which Moesgaard Museum was sent for. There was no doubt that digging had taken place at the site, but only in the upper topsoil. This is very probably because the subsoil here is unusually hard and even with a mechanical excavator digging was difficult. The conclusion reached here was that a person or persons had attempted to dig a grave but had been forced to give up because they did not have the appropriate tools. Subsequently, the ex-husband was, based on circumstantial evidence, convicted of homicide and after serving his sentence will be subject to permanent deportation. The body has still not been found.

In 2009, a woman disappeared from her home in the Aarhus area in what later came to be known as the Lisbet case. A former boyfriend was remanded in custody but shortly afterwards committed suicide in his cell. Despite extensive investigations, a body was never found. Based on new information received in 2017, Moesgaard Museum was asked to assist in the search of a water-retention basin located near a motorway. Despite the exceedingly difficult excavation conditions, involving a more than 1-m-deep layer of mud and silt, the presence of a body could be conclusively dismissed.

In a series of minor investigations, graves have been searched for without any being encountered: One of these cases was in eastern Zealand in 2011. A witness stated that a missing and presumed murdered person had been buried and possibly also dismembered in a wood to the south of Køge. A police dog handler had, prior to his, searched the area without finding any trace. Manual exposure of the area revealed a number of minor disturbances, all of which proved to stem from the dog handler's exploratory excavations in the places where he believed his dog had indicated. No grave was encountered.

In 2014, at an asylum centre in northern Zealand, an asylum seeker confessed to the murder of another asylum seeker and pointed out the place where he allegedly had buried his victim. Even though the perpetrator did not appear particularly credible, the information was nevertheless taken seriously, not least because it was not possible to make contact with the presumed victim. It should however be said that it is not unknown for asylum seekers to disappear from asylum centres. The surface soil in the indicated area was removed with a mechanical excavator, and it very quickly became apparent that no digging had taken place previously in this particular place. The conclusion was reached that the "perpetrator", who was scheduled for deportation due to a number of serious assaults, was attempting to be imprisoned in Denmark, rather than be deported and serve his sentence in his homeland.

Moesgaard Museum's wet-sieving equipment has also been put to good use on several occasions. In a very high-profile case near Odense, known as the double murder in the thousand-year forest, a married couple were completely wantonly gunned down with a machine pistol and further fired upon as they lay on the grass. Danish special forces undertook a metal-detector search of the area, but there was a suspicion that not all the projectiles had been found and Moesgaard Museum was contacted. The area was then searched again with metal detectors, and it quickly became apparent that there were several projectiles buried in the ground. An attempt was made to see if it was possible, through careful removal of the surface soil, to demonstrate the direction of fire for the individual projectiles. However, due to the soil conditions and the dense root mat formed by the grass, this idea had to be abandoned. Instead, the soil was transported to Moesgaard Museum where it was wet-sieved, and the remaining 8 of in all 32 projectiles were recovered. The perpetrator has since been convicted and imprisoned Bang (2014).

The museum's sieving equipment has also been used in arson cases, involving either small amounts of material being transported to Moesgaard or the entire sieving set-up being transported to the scene of the crime. The latter was the case with an arson murder in western Jutland on New Year's Eve 2015–2016. A woman was strangled with an electrical cord, and the house subsequently set on fire and burnt down. About 10 m^3 of material was wet-sieved, and parts of a mobile phone, important for the investigation, were recovered. A man, who incidentally had a previous conviction for murder, was subsequently convicted of the murder.

The last case that will be mentioned here is very recent, from late 2017, but has its roots much further back in time. In the 1940s, a man living in a very hilly area of western Zealand was involved in the resistance movement against the occupying German forces. One of his tasks was to take care of weapon consignments dropped from planes over Denmark by the Allies. At some time in 1944, there was apparently a fear that the Germans would pick up on these illegal activities. As a consequence, a large quantity of weapons, weighing possibly as much as a tonne, was buried in the hills to the east of the house where the family lived. The police focus on the case now, so many years later, is due to the interest of several criminal groups in the story and the risk that the weapons could once again come in circulation if these groups succeed in finding out where they were buried. It is evident from early aerial photographs that at the time the weapons were buried, the hills were largely treeless. With time, however, from the 1950s onwards, they became increasingly covered by woodland. The area today appears completely afforested, and in some places, access is difficult. According to an account passed down via the resistance man's grandchild, there is potentially a large area where the weapons could lie buried. Combined with the difficulty of access, locating the weapons was therefore no easy task. The grandchild had though a couple of supplementary pieces of information. According to what he had been told, from the place where the weapons were

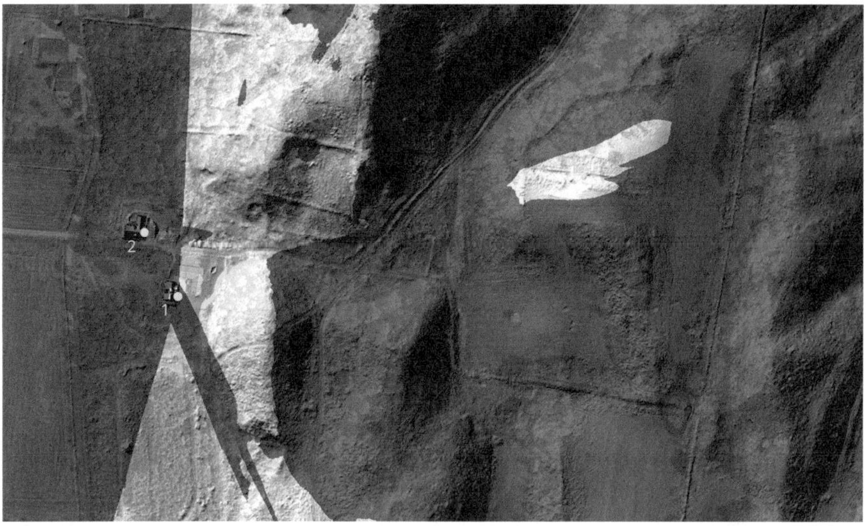

Fig. 11.3 LiDAR scanning of an area where, in 1944, a large quantity of weapons was buried. In principle, the weapons could lie anywhere to the east of the resistance member's house (no. 2 on the plan). According to his grandchild, from the burial site, it was possible to see the gable of the house (no. 2) and the kitchen window of the neighbouring house (no. 1). A viewshed analysis based on this information has limited the potential area considerably. (Peter Jensen, Archaeological IT, Moesgaard Museum/Aarhus University)

buried, it was possible to see both the gable of the family's house and the kitchen window of the neighbouring house. Using this information, a viewshed analysis of the area was performed which limited the potential burial area substantially (Fig. 11.3). A survey was subsequently carried out, based on this analysis, and several areas were identified as being of potential interest. The investigations are still ongoing, so it is unknown whether the initial evaluations were correct.

11.8 The Future

In the only review of forensic archaeology in Denmark published to date, the previously mentioned article by Jørkov and Lynnerup (2015), it is stated in the conclusion that "it is difficult to see a more substantial use of forensic archaeology in Danish casework, and probably not in terms of archaeologist specialising in forensics". As is evident from the above, increasing use is being made of archaeologists by the police, not just in connection with the excavation of graves but also in other types of criminal investigations. And even though it is generally said in Denmark that *it is difficult to make predictions, especially about the future*, I nevertheless believe that forensic archaeology is here to stay – also in a Danish context. It is also to be expected that archaeological methods can be applied in other aspects of police work. In particular, there appears to be a potential in relation to buried narcotics and weapons (terror), both of which are offences carrying a range of sentences comparable with those for homicide. Similarly, the author believes that excavation with archaeological methods will be able to make a positive contribution to the detection work in certain types of arson cases, especially those where everything has been burnt to the ground. That said, it should be made clear that the organisation which has been built up around forensic archaeology in Denmark is a fragile construction based primarily on personal contacts, and a serious lack of institutional anchorage is still felt. In this respect, work in progress under the auspices of the ENFSI to produce a Best Practice Manual and the possibility of obtaining accreditation as a forensic archaeologist will undoubtedly help.

Chapter 12
Recent Recruitment of Forensic Anthropologist and Odontologist in Carabinieri Scientific Investigation: Multidisciplinary Approach and Scene Investigation in Case of Hidden Corpse

Chantal Milani and Carlo Belardo

Abstract The roles of the forensic anthropologist and odontologist, individually, are quite rare in Italy, and there are a few situations in which these specialties operate successfully to be of significant benefit to the investigations. These two important roles are treated in this chapter, for introducing the reader to the problem.

Keywords Forensic anthropology · Forensic odontology · DVI · identification of unknown bodies

12.1 Introduction

Forensic sciences have reached a high level of specialization, where it becomes essential to create interdisciplinary teams of professionals working side by side and in synergy, in order to respond to the issues thrown up by the case. Outside Italy, this type of reality is a common and consolidated practice, constructive not only for the case itself but also for respective professional development.

Today, everywhere, it becomes a mandatory to provide the judicial authority with answers to increasingly complex questions, if we want to keep up with the times and with the progress of science. In 2016, in Italy, the Carabinieri recruited a forensic odontologist and anthropologist as a RIS officer in Rome, the only such case till date in Italian law enforcement, working in the field of scientific investigations. This

C. Milani (✉)
Carabinieri R.I.S. Scientific Investigation Department, Rome, Italy
e-mail: chantal.milani@carabinieri.it

C. Belardo
Crime Scene Investigation Unit – Carabinieri R.I.S. Scientific Investigation Department, Rome, Italy

P. M. Barone, W. J. M. Groen (eds.), *Multidisciplinary Approaches to Forensic Archaeology*, Soil Forensics, https://doi.org/10.1007/978-3-319-94397-8_12

step, which belongs to two highly specialized and complementary disciplines, has proved useful in various investigations and in the identification of the victims of the earthquake that hit Central Italy in the summer of 2016.

12.2 Forensic Anthropologist and Odontologist in Italy

The figures of the forensic anthropologist and odontologist, also taken separately, are not so common in Italy, and there are few situations in which these specialties operate successfully to bring significant benefits to the investigations.

Corpses of different types may be discovered at the crime scene, including charred ones, bodies in an advanced state of decomposition, those recovered from the water, found in open fields, mutilated or even skeletal remains hidden for a long time, where most of the information have to be drawn from the bones and teeth.

In these cases, the presence of the medical examiner alone is not enough, and there is the need to employ specialists who work together right from the inspection, along with the other figures operating on the scene to recover, preserve, and consolidate elements essential for the continuation of the investigations and subsequently to complete the analysis of the remains found. In most of cases, losing the information provided by the first inspection or missing small findings (e.g., little bones or dental fragments or prosthetics) means that some answers are irretrievably lost.

The forensic world has been receiving dramatically increased attention in recent years due to the substantial impact that the airing of legal cases and the various TV serials have had on the public – the so-called CSI effect. All this has raised expectations, not always in a realistic way, since the answers provided by the various forensic sciences are subject to a statistical interpretation, and it is not so easy to explain with a realistic perception. Despite this, all this attention has placed the forensics sector under the spotlight and increased its appeal.

In Italy, forensic anthropology and odontology are often confused with forensic pathology and with physical anthropology in the field of bioarchaeological and museum anthropology and with clinical dentistry. In reality, very few physical anthropologists and odontologists can be called "forensic," since this term is often misused. Much confusion arises from the fact that these disciplines are prevalent in North America, where training is very different, right from the university study programs.

The forensic field inevitably requires specific and specialized knowledge in addition to the scientific knowledge acquired in the underlying subject that includes notions of law and criminal procedure, and, with regard to forensic anthropology and odontology, there must be specific training targeted at highly compromised or decomposed human remains, related to modern contexts. This cannot be acquired through courses lasting a few weeks. This is because the field in which one is going to work has important rules and legal issues as well as technical needs, and some activities could take place on bodies that are not entirely skeletonized, which may require a radiological approach and knowledge of biomechanical elements, even if one has to investigate only the skeleton and the teeth. Therefore, concerning the training available in Italy, it is advisable that these figures acquire a medical (medical doctor or odontologist) and not merely a biological or bioarchaeological background.In the specific case of the forensic dentist, he or she must also be enrolled in the professional register of dentists.

The main sectors in which the forensic odontologist-anthropologist dual figure can be successfully employed include:

- Identification of the body. In Italy, there are more than 800 unknown corpses, excluding those recovered at sea due to migratory flows and 47,946 missing people between 1974 and 2017.
- Age assessment of unaccompanied minors, necessitated by the large number who arrives in Italy without reliable documents.
- The anthroposomatic analysis of subjects portrayed in video surveillance to be compared with the body and facial features of subjects suspected by the law enforcement.
- The analysis of bite marks sometimes associated with maltreatment, abuse, and neglect, with evidence of skeletal or non-skeletal injuries.

12.3 Inspections of Human Remains Discovered: Importance of the Multidisciplinary Approach

The environments linked to a suspicious death are manifold: closed room, open room, aquatic areas, woodlands, burials, etc. This environmental variability is associated with several conditions of the corpse: total or partial skeletonization, maceration, deactivation, carbonization, mummification, or merely the advancement of decomposition processes, due to which the analysis and identification of the body can be complicated. The presence of the forensic anthropologist and odontologist can be useful right from the on-site inspection stage, whenever bones and teeth can potentially provide information on the identity of the body and the dynamics of the event, and therefore a correct and complete (as far as possible) recovery of the remains and anything associated with them is essential.

Law enforcement, along with specialized crime scene investigation units, arrives on a scene starting from the first moments, which is why it is important to be able to organize a proper team of specialists capable of dealing with the specific case. A close-knit team of professionals makes for easy workflow, interaction, and communication throughout the whole case management.

Preliminarily, the scene often involves the search for informants who generically reveal the corpse that has been hidden for a long time and present in a more or less extensive area. In this case, it is critical to set up multidisciplinary teams, which, in addition to the forensic anthropologist and odontologist, could involve also the collaboration of law enforcement canine units (K-9) specialized in the search for cadavers and traces of blood, as well as forensic geologists and forensic archaeologists. The first aim is to restrict the area of intervention as much as possible, for a correct excavation that brings to light and preserves the remains by recovering them as thoroughly as possible. The “cadaver dog” is an extremely useful resource. The Carabinieri have a dedicated canine unit located in Bologna that operates throughout the country.

The human remains follow the chain of custody from the complete recovery, inventory, collection, and related documentation of the scene to the final laboratory stages to complete the traditional autopsy or skeletal and dental analysis. The forensic anthropological analysis (ancestry, sex, age, stature, etc.) of cadavers (although in bad condition) can also lead to the assessment of cause of death, the dynamics of the event, the research for a subject’s identity in the missing person database compared with the anthropological profile, dental aspects and/or DNA, and anything else useful for legal purposes. Since 2016, the Carabinieri RIS Rome dealt with several cases that required the intervention of the forensic anthropologist and odontologist, including the identification of the victims (DVI) of the earthquake that struck some cities of Central Italy on the night of August 24, 2016. Amatrice and Accumoli were the two most affected cities. Two hundred and forty-six victims were identified in a matter of a few days, thanks to the teamwork and the multidisciplinary approach, despite the emergency situation and a difficult logistics situation complicated by the continuous aftershocks (seismic swarms).

Chapter 13
Forensic Archaeology in Russia: Past Developments and Future Approaches

Tatyana Shvedchikova

Abstract Forensic archaeology in the Russian Federation faces several problems on its way to being established as a discipline. The past developments in the fields connected to forensic archaeology such as taphonomy and time since death estimation provide the theoretical and experimental basis of the discipline's application. The key problems and possible ways of development are shown.

Keywords Taphonomy · Crime scene investigation · Time since burial estimation · Historical owerview

As it is shown in previous publications, forensic archaeology as an independent discipline does not exist in the Russian Federation (Abramov et al. 2015). Archaeological approaches are usually ignored by the police force during forensic cases' investigation. The reasons behind this predicament could be divided into several parts.

First, it is the strict regulations and sometimes miscommunication between the organizations which have the right to deal with crimes on different levels (Police, Investigative Committee, Federal Security Service, special departments in the Ministry of Justice, the Ministry of Defense, and so on). On some levels, the involvement of third (civilian) parties in the investigation process becomes impossible or becomes an extremely bureaucratic and long drawn-out procedure that extends the investigation time.

Second are the special aspects of federal laws, which divide the areas of responsibility between different services that deal with interments. If a burial is older than 100 years, it is an archaeological object. All actions pertaining to it are regulated by special law and the Ministry of Culture; its investigation is conducted by professional archaeologists with special license. If the burial is dated less than 30 years, it leads to the initiation of a criminal case. Its investigation is handed over to the Police or Federal Security Service, and legal control is placed with the Ministry of Justice.

T. Shvedchikova (✉)
The Institute of Archaeology Russian Academy of Sciences, Moscow, Russia

P. M. Barone, W. J. M. Groen (eds.), *Multidisciplinary Approaches to Forensic Archaeology*, Soil Forensics, https://doi.org/10.1007/978-3-319-94397-8_13

We have a time span of 70 years between the 30 and 100 years, which is not so strictly regulated. There were two bewildering cases where the suspect took part in the serious crime, the crime detection had no time limit, and the interment became the forensic case. However, the majority of revealed cadavers belong to the WWI and WWII victims. The investigation of cases like this is formally regulated by the Ministry of Defense, and the execution is on the nongovernmental community-based organizations, which were united under the umbrella of all-Russian social movement "Exploratory Movement of Russia" in 2013. The necessity of forensic archaeology being established as a discipline is perceived without understanding. For professional archaeologists, there is no scientific interest in fresh cadavers and their search. The rich archaeological history of the country proposes the study of different epochs and topics without harsh competition. People who deal with WWII victims ask questions such as "why should we follow the archaeological methods?" Our main purpose is fast identification. And if in this case, the situation is gradually changing with time over a 100-year period, which already made WWI the field of archaeological regulation, the complete incomprehension of archaeological approaches necessarily comes from the law enforcement community. There is no conception about the quantity of possible information the researcher can obtain by applying the archaeological methods of study. The high-profile murder case of the Romanovs Royal family, which drew attention during the last couple of years, made experts discuss the initial stage of remains' recovery and not appropriate actions during the excavation process.

So, at this moment, the good examples of the application of the archaeological method are sporadic and based on the personal initiative of concrete criminal investigators. This tendency does exist, but does not, as yet, have a regular basis and strict legal frames. Some cases lie on the border with the regular archaeological case. During the last few years, its extensive use has got the identification of historical characters, which are usually excluded from classical archaeological study. In "General Ermolov's court case," archaeology helped in proving the absence of legal components of a crime (Abramov et al. 2015; Engovatova et al. 2014). Two workers, who performed the repair of the flooring in Trinity Church, where the family crypt of Ermolov's family is located, were accused of the desecration of the grave of a well-known military leader of the nineteenth century. Artifacts confirmed the four stages of penetration into the crypt. The first stage was performed in the 1930s of the twentieth century (coins and newspaper fragments), the second was carried out during WWII, when the church was used as a horse barn in the occupation of the city (the "Bosch" automobile headlight), the third was in the 1970s, during the church reconstruction of the Olympic Games, and the last in the beginning of 2000s (sweetie wrapper and the bank note fragment). The accused persons turned out to be innocent, and all charges filed against the defendants were dropped due to the absence of criminal elements.

If the methods of archaeological investigation of forensic cases were not used, another very important question connected to the decay of cadavers was developed since the early 1940s of the twentieth century in the frames of the taphonomical approach. Every paper, discussing the processes' flow with corpses after death and burial, could not do without mentioning Soviet scientist Ivan Efremov as the founder of the discipline of "taphonomy." This is a paradox, but on its motherland, the new approach to the study of human remains and burial complex, which is being very intensively developed nowadays mostly due to experimental facilities, did not develop into a scientific school as it did in most European countries and America. The term "taphonomy" has recently been used for the specific topic of paleontology, but did not migrate to physical anthropology or archaeology and forensic medicine. Although rapidly developed in the 1970s in the forensic medicine school of PMI, the definition did not perceive this theoretical concept.

Everett C. Olson said, "scientist, philosopher, writer and Russian to the core, Professor Ivan Antonovich Efremov seemed to me a unique phenomenon in a society where such a free, imaginative spirit had no business to be…." Ivan Efremov was a brilliant Russian scientist who dedicated more than 200 publications to the different aspects of study of the geological past of our planet.

His main idea was to unite geological chronicle and biology, return to the prehistoric times and tell the colorful story of living organisms from their birth to death, watch their decay and fossilization, and finally show this picture to us.

Efremov was born in 1908 in a small village near Saint Petersburg. The beginning of the twentieth century was a hard period of the revolution, and big changes were taking place in Russia. At the age of 12, Ivan joined the Red Army unit during the civil war as a "son of regiment." He performed various tasks and spent 1 year in the sea as navigator. Obstacles hardened him but did not kill his thirst for knowledge. Coming back to Saint Petersburg and after finishing school, he decided to become a paleontologist. He had read an educational article about paleontology by academician P. Sushkin. He was so excited that he wrote him a letter.

Efremov began to work as his laboratory assistant in the geological museum. From 1925, he took part in a variety of paleontological expeditions in the Urals, Central Asia, and Volga region. At the same time, he studied biology at Moscow State University.

During the expeditions, Efremov uncovered the mass graves of the fossilized skeletons of various dinosaurs. It made him think about the conditions of remains' appearance and the way they fossilized. He named Charles Darwin and Sushkin among those who were the predecessors of taphonomy and:

> …I propose for this part of paleontology the name of "TAPHONOMY," the science of the laws of embedding. … It stands on the border of paleontology, uniting it both with geology and biology into one general bio-geological historical method of study. (Efremov 1940a)

In 1940, he published, in English, the well-known article in *Pan-American Geologist* and an article in Russian, where he named the discipline "taphonomy." The scientific public did not immediately accept the new discipline (Efremov 1940b). By 1943, he had already written the book *Taphonomy and Geological*

Chronicle, but it was published only in 1950. Ivan Antonovich wrote in the preface to the book that the delay was due to conditions of wartime and Efremov being busy with other works. Later, he realized that most of the statements of the new discipline were considered heretical, and the book had not been published for a long time in order to not ruin the basic foundations of paleontological science. Ivan Antonovich also believed that the discipline was born too early and would be in vigor in just 20 years, like it happened.

In this book, besides mentioning the characteristic of mass graves of fossil remains, he stated the main principles of taphonomic research. He defined the main factors of fossilization of organic remains and the states in which different agents would influence the way the fossilization process progresses.

Among them are the preservation state of remains at the moment of burial, physical, chemical, and biochemical soil properties (and in this case, of course, the climate of the area) and secondary disturbance of primary burial (Efremov 1950).

From the taphonomic point of view, he examined the findings of prehistoric human remains and concluded that at that times, the diversity of forms was larger than we suppose. So, the taphonomical approaches according to the author's opinion are applicable to the study of human remains.

For this fundamental research and his book, Efremov was awarded the Stalin Prize in 1952 (the highest honor for prominent achievements in the field of science, literature, arts, and architecture). Some of his colleagues say that he was not only honored for the developments in paleontology but also for the prominent work in geology. He took part in geological expeditions, seeking the uranic ore, mercury, oil, copper, gold, and other strategic feedstocks.

In the twentieth century, the taphonomical studies were developed in the frames of paleoecological approach, mostly in paleontology. Archaeologist Yuri Smirnov in his book *Labyrinth: Morphology of Intentional Burial* mostly discussed theoretical questions on necrology and taphology (a term proposed to describe the knowledge about the ways of intentional burial of human remains) from the ancient to the modern times (Smirnov 1997). He also mentioned that, unfortunately, nobody applies taphonomical approach in Russian archaeological and anthropological science.

Just in forensic medicine, we can find publications regarding the decay of soft tissues by professor Rubezhansky, which were based on large experimental data (Rubezhansky 1978). These studies were connected to the time since death estimation investigations. In his dissertation and numerous publications Anatoly Rubezhansky shed light on the way the human body decayed in different conditions and various environments. In his works, Rubezhansky noted the importance of paying special attention to the methods of registration and the skeletonized material (found in the open air, in the water, or buried). Bones should not be the only characteristic in the limelight; environmental conditions (flora, fauna, soil characteristics, depth of burial) should also be registered and sampled (Rubezhansky 1975). The Department of Forensic Medicine at the Dnepropetrovsk Medical Institute, which started during the 1970s, became the leading school in the study of the process of decomposition of the human body. Different aspects were studied: the mineralization process of human bones in the soil, bone weathering in different burial conditions, spectral analysis of teeth and bone which was used in reference to time since death estimation, the change and decay of bone protein, generalized model of human body decomposition, and its particularities in case of different soil conditions. Experimental material included the corpses buried from 2 to 38 years ago. As a result, in 1978, the complex methodical book *Estimation of the Time Since Burial on the Skeletonized Material* was published, which remains to date the main guidance for time since death estimation in the post-Soviet countries (Rubezhansky 1978).

Three hundred corpses buried in three different soil conditions were examined and numerous methods were applied. Among them were ultrasound bone decalcination, bone staining for collagen detection, visual macro and microscopical examination (color, erosion rate, bone density estimation), spectral analysis, and soil analysis (color, mechanical composition, structure analysis: humus content, pH level, soil moisture, microbiology). The main disadvantage of the performed work was that the study only included the several types of soil available in the region for experiments. Unfortunately, the investigation did not continue in the same manner since the 1990s of the twentieth century. The absence of complex studies at the same level for the other soil types forced the experts to adopt methodical conclusions to the local situation, which could drastically differ from the described condition.

Rubezhansky's research took into account a number of previous studies (Orfila and Lesueur 1931; Walcher 1950; Foerster and Goldbuch 1954; Kamps and Parchese 1956; Specht and Berg 1958; Prokop 1960; Scheibe et al. 1962; Berg 1964; Neckermann 1958 cit. ex Rubezhansky 1978). Observations were conducted for three different types of soil: (1) leached low-humic deep black earth soil, (2) humus–carbonate soil on the marl rock and mountain humus–carbonate soil, and (3) carbonated low-humic deep black soil. More than 300 corpses were studied, and contemporary and subfossil samples were involved.

13.1 Leached Low-Humic Deep Black Earth Soil

Total decomposition of the body comes up in 2–3 years. The bone morphology is not changed and only starts to change in the 20th year after burial. There is almost no difference in color, as changes begin in the 25th year. The bones of individuals buried in coffins sharply differ from those who were buried in simple pits. The signs of erosion are absent. Differences in spectral analysis (Si, Al, Mg, and Mn rates are increasing) and coefficients Ca/Al and Ca/Si, Ca/Mg, and Ca/Mn allow the estimation of the time since burial. Differences in Al accumulation are detected for adult and subadult persons. The speed of decalcification process allows the estimation of the time since burial ±2 years.

13.2 Humus-Carbonate Soil on the Marl Rock and Mountain Humus-Carbonate Soil

Bones are solid and dry, polished by 18–20 years of burial. Small cracks begin to appear in 31 years. Colors depend on the time of burial; after the 21st year, brown stains appear. Soft tissues could be found on the skeletons up to 11 years since interment. Erosion signs are rare. Differences in spectral analysis (Si and Al are increasing) and coefficients Ca/Ti allow the estimation of the time since burial. The speed of the decalcination process is faster than the bones buried in leached low-humic soils.

13.3 Carbonated Low-Humic Deep Black Soil

Bones are solid and dry, polished by 18–20 years of burial. Small cracks begin to appear in 31 years. Colors depend on the time since burial; after the 21st year, brown stains appear. Differences in spectral analysis (Si and Al are increasing) and coefficients Ca/Al and Ca/Si allow the estimation of the time since burial. The speed of the decalcination process is faster than in bones buried in leached low-humic soils.

Subfossil and fossil archaeological material does not show the decrease of protein component with time, revealing a little lower characteristic when compared with fresh and buried 38 years since death material.

It is clearly shown that bone samples should be necessarily followed by soil samples and studied in the complex for drawing correct conclusions.

The last chapter of the book is dedicated to recommendations, which we can link to the recommendations provided in the frames of forensic archaeology nowadays. The importance of careful excavation of clandestine burials by avoiding the use of any machinery and paying attention to the environmental circumstances is stressed. It is highly recommended to dig up bones using archaeological methods and paying attention to the context and fixation of the excavation process (including photofixation). Soil samples (with a control sample 15–20 m away from the grave) should be taken. The body position is necessary to describe the flora and fauna inside the pit. Special attention should be payed to the scattered bones, collective burials, and individual identification in this case. The expert should take all artifacts and follow the protocol of laboratory investigation.

As we see, the practical recommendations fit the criteria of forensic archaeology prescribed for the field work. The problem of recommendations advisory, no obligatory character, and lack of forensic experts in the field made them nonexecutable in mass.

During the last 20 years, few papers appeared displaying how taphonomical methods are applied during archaeological or forensic investigation. Most of them are connected to the French school of archaeothanatology of Henri Duday and the joint Russian–French archaeological project held during the 2000s in the Ural region.

Future steps in forensic archaeology in Russia are connected to the promotion of the discipline's importance and the quantity of information it can provide. Its inclusion in the legal system will not be easy and appears to be very individual in every region of the country, depending on the local specificity (geographical area and population density) and the complexity of legal regulation. Every perspective in this sense could be a collaboration between services dealing with WWII search and scientific society.

Chapter 14
Operation Nobility: The Identification of a Missing Soldier from the Battle of Arnhem, 1944

Capt. Geert Jonker

Abstract The Recovery and Identification Unit of the Royal Netherlands Army (RIU) was established in January 1945 under the Dutch Free Forces of the Interior, in the liberated south of the Netherlands. On behalf of the Netherlands Ministry of Defence, RIU (part of the Royal Logistic Corps) is responsible for location, exhumation and identification of missing World War II victims, both civilian and military, regardless of their background or nationality. Its main objective is to help next of kin achieve closure and ensure a reinterment in a marked grave, in accordance with national and international conventions, regulations and legislation. The Netherlands government continues to consider this a moral obligation, a duty of care and a debt of honour. The work of the unit combines the fields of forensic and conflict archaeology, physical anthropology (i.e. osteology and odontology) and military history. Basically RIU matches biological profiles to medical profiles and historical profiles. The unit is based at the Dumoulin barracks at Soesterberg, the Netherlands, where it works from a human osteology laboratory. The purpose and intent of this investigation was to identify the skeletal remains of a supposedly World War II British soldier, found on 6 February 2013 in south Arnhem, the Netherlands. The answers were obtained through the following types of research: forensic archaeology, human osteology, human odontology, study of the recovered artefacts, desktop research, genealogy and DNA analyses.

Keywords Battle · Arnhem · Oosterbeek · 1944 · Market-Garden · Conflict archaeology · Identification · Osteology · RIU · Wiltshire Regiment · DNA · Dutch

Capt. G. Jonker (✉)
The Recovery & Identification Unit of the Royal Netherlands Army (RIU), Soesterberg, The Netherlands
e-mail: gbk.jonker@mindef.nl

P. M. Barone, W. J. M. Groen (eds.), *Multidisciplinary Approaches to Forensic Archaeology*, Soil Forensics, https://doi.org/10.1007/978-3-319-94397-8_14

14.1 Introduction

On 6 February 2013, RIU received a telephone call from the KWS Special Works Company, a civilian unexploded ordnance clearance company. Personnel from this company were working in the area between south Arnhem and the town of Elst, which was a military theatre of operations during the aftermath of Operation Market Garden in September–October 1944. During their work they had come across a clandestine human burial. Because of the military history of the area, the work was subsequently stopped and the police as well as RIU were summoned to the site.

It was a cold, sunny day with blue skies, although there was still some snow on the ground. The soil was of a clayish type, the area lying right between the rivers Rhine and Waal, fertile, marshy land which is known to have been occupied by man for thousands of years. After arriving at the site, two sets of skeletal remains (SR) were uncovered, lying side by side in a collective field grave. The skeletal remains were registered as 006/13 and 007/13. SR 006/13 was found face down at a depth of approximately 80 cm, on the southern end of this collective grave. Apparently his head had been covered prior to the burial. SR 007/13 was found buried on his back, at the same depth. Some military items, i.a. an entrenching tool, a water bottle and two Bren magazines in a pouch, were recovered with the SR. Also a Wiltshire Regiment cap badge and a number of great coat buttons were recovered with SR 006/13. Operation Nobility focuses on SR 006/13 (Photos 14.1, 14.2 and 14.3).

Photo 14.1 Yellow rectangle indicates location of the field grave. (Source: Google Earth). GPS location: N 51°56′27.8″ E 5°50′49.9″. Map reference: 31 U FT 95699 58311

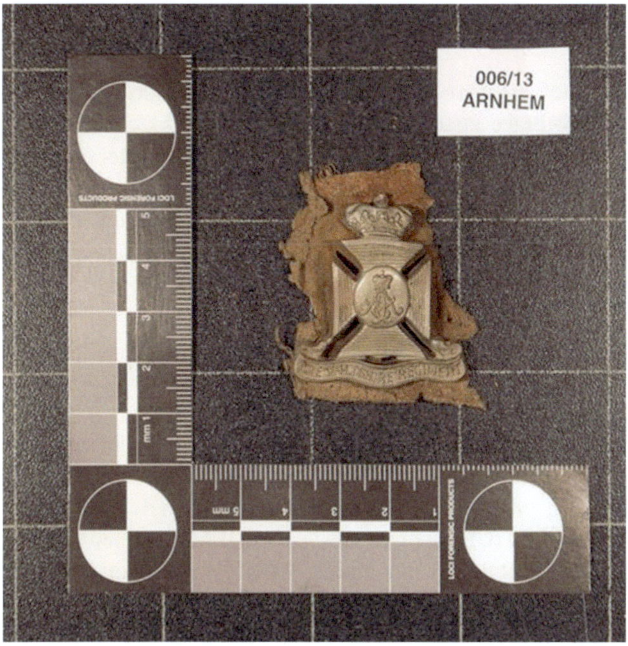

Photo 14.2 Wiltshire Regiment cap badge found with SR 006/13. (Photo: RIU)

Photo 14.3 SR 006/13 (south, face down) and SR 007/13 (north, face up) in field grave. (Photo: RIU)

14.2 Artefacts Found with SR 006/13

The following artefacts were found accompanying SR 006/13: webbing belt M37; cross straps M37; brace attachments M37; L-straps M37; ammo (basic) pouches (set) M37; entrenching tool 1908 pattern; two-pint water bottle and carrier; unidentified leather pouch to be carried on web belt; army ankle boots and anklets; battle dress 1940 pattern; great coat, wool, 1940 pattern; and cap badge Wiltshire Regiment, plastic, economy issue. No personal belongings or weapons were found.

14.3 Physical Anthropological Research

SR 006/13 and SR 007/13 were transferred to the RIU's laboratory at Soesterberg. Physical anthropological research led to the following conclusions: preservation, fair; completeness, incomplete; taphonomical stage 4 (Behrensmeyer 1978); and decomposition state 10 (Clarke et al. 1997). The skeletal remains were in fair condition yet heavily decalcified and incomplete. This is possibly due to the fact that local farmers have used manure as a fertiliser on the field for decades, making the soil highly acidic.

14.4 Biological Profile

Sex:	Male, Caucasian
Age estimation:	17–25 years of age (Meidl and Lovejoy 1985; Suchey and Brooks 1990; Todd's 1920; Buikstra and Ubelaker 1994; William 1995; Owings Webb and Suchey 1985)
Stature in cm's:	170,4 cm (Trotter 1970, standard error ±3,27 cm
Anomalies:	Severe ballistic trauma to the skull, completely fragmented, parietals missing

There was a severe (high velocity/ballistic) trauma to the cranium, and both parietals were missing. The nasal bone was angled in an upward position. This bone trauma was most probably due to the explosion of a mortar bomb or artillery shell. There were no other signs of trauma or disease. The trauma to the skull was most probably the reason why the individual's head was found covered prior to burial: It must have been an unpleasant sight for the burial party (Photos 14.4 and 14.5).

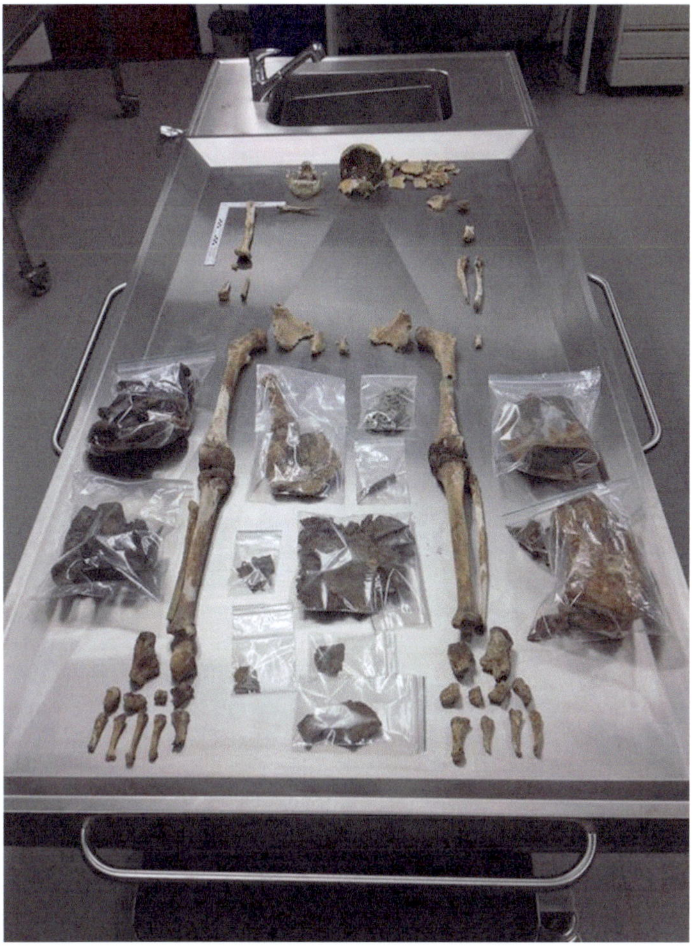

Photo 14.4 SR 006/13 in the laboratory. (Photo: RIU)

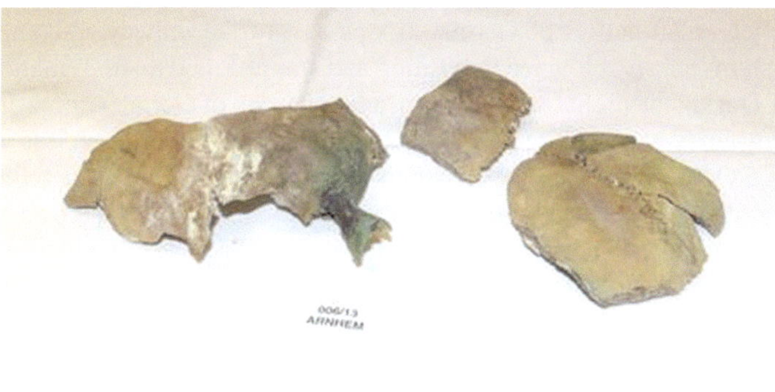

Photo 14.5 Cranium (high velocity/ballistic trauma). (Photo: RIU)

14.5 Desktop Research

14.5.1 4th Battalion, the Wiltshire Regiment

As part of 129th Brigade (43rd Wessex Division), 4th Battalion, the British Wiltshire Regiment participated in the Battle of Normandy, landing in France on 24 June 1944. During the Normandy Campaign, the Wiltshires were involved in all major battles: the Odon Box, the battle for Hill 112, (Operation Jupiter, the battle for Caen), the capture of Mont Pinçon and the crossing of the river Seine at Vernon. At the start of Operation Market Garden (17 September 1944), the Wiltshires were in Helchteren, Belgium, just across the Dutch border. The 43rd Wessex Division was part of 30 British Corps, who were supposed to go up the corridor (so-called Hell's Highway) and advance into Holland in order to relieve the Airborne troops who had taken the bridges over the major rivers, Arnhem being the most northern one. 4th Battalion crossed the Dutch border on 22 September 1944, moving north via St. Oedenrode, Uden, and finally to the town of Grave, where they arrived at approximately 2:00 AM. They did not get much rest; the battalion moved towards Elst, just south of Arnhem, 6 h later. Upon arriving at Elst, they encountered German opposition during which A company sustained heavy casualties. During the following days, the battalion held their positions near Elst. On 29 September 4th Battalion Somerset Light Infantry was holding Elst, with 5th Battalion, the Wiltshire Regiment, to their north. 4th Battalion was in reserve near Valburg. On 1 October the German attacks concentrated on the 129th Brigade front with the Somersets fighting off all attacks on Elst. The level crossing 2 miles north of Elst was of highly strategic importance and heavily fought over. On 2 October C company, 4th Battalion, was under the command of 5th Battalion. They stopped German troops from infiltrating between the positions of 4th Battalion Somerset Light Infantry and 5th Battalion, the Wiltshire Regiment, just north of Elst, along the railway line, in the area of De Laar farm (Photo 14.6).

Twenty-four hours later, during the night, 4th Battalion relieved 5th Battalion in their forward positions in the De Laar area and level crossing. Whilst doing so, D company sustained casualties from the constant shelling. On 4 October D company also engaged small parties of German troops east of the railway. C company, 5th Battalion, temporarily remained in position and was attached to 4th Battalion for the next 48 h. During the night (4–5 October), a large group of German troops infiltrated D company lines but were successfully dealt with, and 50 prisoners of war were taken. On 5 October an effort was made by D company, 4th Battalion, to push back the enemy who still held on to the area west of the railway near the De Laar farm. In the late evening of 5 October, at approximately 9:00 PM, the relief of the battalion, by 501 Parachute Infantry Regiment of the 101st (US) Airborne Division, commenced. The battalion moved the companies, one by one, to a forward assembly area (an area northeast of Oosterhout) and from there to the brigade concentration area (area Weurt/Beuningen). The move was completed by 5:00 AM on 6 October (Photo 14.7).

Photo 14.6 De Laar farm in 1940

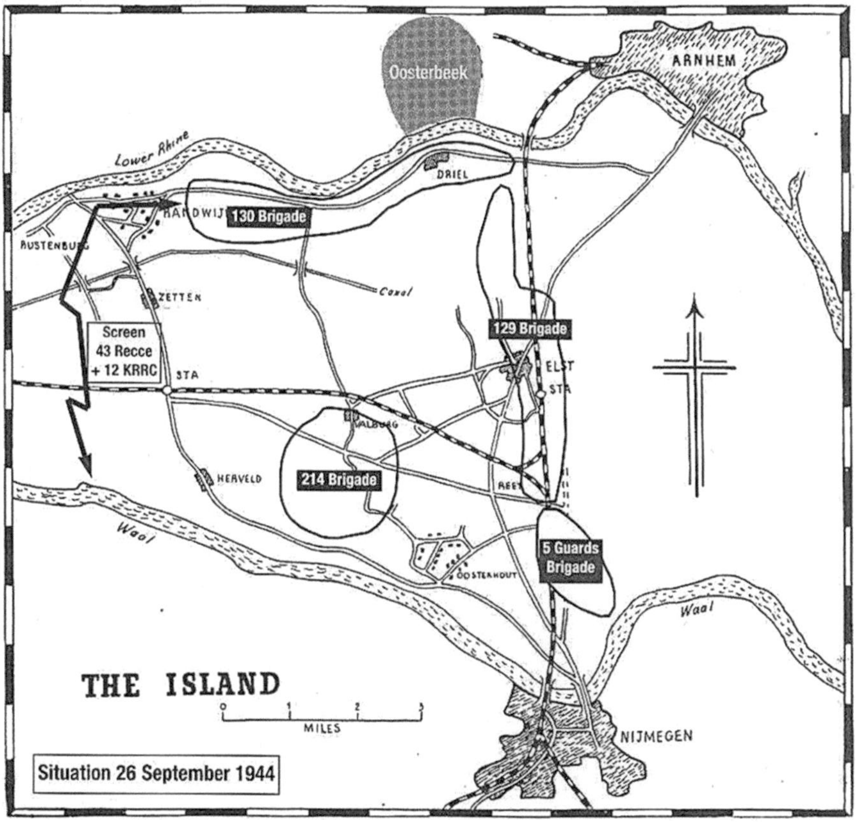

Photo 14.7 The Island, area between Arnhem and Nijmegen

Based on the historical research, these two soldiers most probably had been killed in action during the early days (1–5) of October 1944 and had served with 4th Battalion, the Wiltshire Regiment. These two individuals had clearly been temporarily buried with the intention of transferring them to a military cemetery at a later stage. But because the area was flooded by German engineers on 3 December 1944, in order to force the allied troops off the Island, the improvised grave marker with which it would have been marked had washed off. And after drying up in the spring of 1945, the entire area was left covered in a thick layer of silt and mud, hence making it impossible for the British Army Grave Registration Units to locate all temporary burials, despite of their locations having been properly registered at the time of burial. The field grave was found on the south-west corner of a small wood which was already there during the battle and will probably have acted as a landmark.

14.6 Candidates

The location of this common grave was in the 1944 4th Battalion, the Wiltshire Regiment area. After researching the 4th Wilts Battalion Roll for this period, a list of seven candidates for SR 006/13 and 007/13 was established:

Service No.	Rank	Name	Age	DoD	Coy	Enlisted into: (Regtl prefix)
14591941	Pte	LEWIS, Harold James	19	03.10.1944	C	General Service Corps
5248907	Cpl	BILL, William Garnett	34	04.10.1944	D	Worcester Regiment
5350463	Cpl	HOLMES, William Henry	22	04.10.1944	D	Berkshire Regiment
5110483	L/Cpl	NOBLE, Donald Stabler	22	04.10.1944	D	Royal Warwickshire Regt
14204645	Pte	YOUNG, Gilbert Ernest	21	04.10.1944	D	General Service Corps
5126228	Pte	DANGERFIELD, John Thomas	25	05.10.1944	D	Royal Warwickshire Regt
5573297	Pte	EDDY, William John	25	05.10.1944	A	Wiltshire Regiment

The UK Ministry of Defence, Joint Casualty and Compassionate Centre (MOD/JCCC), were asked to provide service and medical records for these seven missing Wiltshire Regiment soldier. Dental records were available for Private William John Eddy and Private John Thomas Dangerfield. Both could be eliminated through their dentals. Other candidates could be excluded based on non-matching height and/or age.

Service No.	Rank	Name	Age	Height (cm)	Coy	1st enlistment	
14591941	Pte	LEWIS, Harold James	19	167,6	C	General Service Corps	
5248907	Cpl	BILL, William Garnett	34	168,9	D	Worcester Regiment	Age
5350463	Cpl	HOLMES, William Henry	22	181,6	D	Berkshire Regiment	Height
5110483	L/Cpl	NOBLE, Donald Stabler	22	170,8	D	Royal Warwickshire Regt	
14204645	Pte	YOUNG, Gilbert Ernest	21	174,6	D	General Service Corps	Height
5126228	Pte	DANGERFIELD, John Thomas	25	179,4	D	Royal Warwickshire Regt	Dentals
5573297	Pte	EDDY, William John	25	177,2	A	Wiltshire Regiment	Dentals

Only two candidates fitted the established biological profiles for 006/13 and 007/13:

- Private Harold James LEWIS
- Lance Corporal Donald Stabler NOBLE

But sadly there were no dental records available for neither Harold Lewis nor Donald Noble. SR 006/13 turned out to have replaced the generic brass "General List" buttons of his great coat (with which it would have come from the manufacturer) with Regimental Buttons, which after careful cleaning turned out to be of the Royal Warwickshire Regiment. Lance Corporal Donald Stabler Noble had originally enlisted into the Royal Warwickshire Regiment. Because of this, as well as a perfectly fitting biological profile (age and height), the focus of this investigation became to investigate whether SR 006/13 could belong to Donald Stabler Noble. Since there were no dental records available and there were no other means of identification, next of kin needed to be traced in order to do DNA research.

14.7 Donald Stabler Noble

Donald Noble was born on 22 March 1922 in the parish of West Leeds, Yorkshire. His mother was Miss Dorothy May Noble. The name of his father was not registered. Complexion, fresh; eyes, grey blue; hair, fair; stature, 1708 cm; religious denomination, Church of England. Addresses in his service records:

- 60 Eden Crescent, Headingley Park Estate, Leeds, Yorkshire
- 60 Eden Crescent, Burley, Leeds 4, Yorkshire
- 60 Eden Crescent, Kirkstall Hill, Burley, Leeds 4, Yorkshire
- 8 Langdale Terrace, Headingley, Leeds 6

NOBLE, L. Cpl. DONALD STABLER, 5110483. 4th Bn. The Wiltshire Regt. 4th October, 1944. Age 22. Son of Mrs. D. Noble, of Burley, Leeds, Yorkshire. Panel 5.

Source: http://www.cwgc.org/find-war-dead/casualty/2659402/

According to his service records, Donald Stabler Noble enlisted into 2nd Battalion, the Warwickshire Regiment in London, and was attested as a boy (employed as a clerk in the orderly room) on 26 June 1938 at the age of 16. In an annual report from September 1938 is stated: "Employed as orderly room clerk, is doing very well, clean and intelligent." Enlisting into the army at the age of 16 most probably was an escape from poverty, being the son of a single mother and without a registered father.

On 22 September 1939, he enlisted into the regular army as a Private, and in November 1940, Donald Noble was transferred to 70th Battalion (young soldiers), the Royal Berkshire Regiment. He was appointed Unpaid Lance Corporal (U/L/Cpl) on 20 January 1941. Three months later he was posted to "Y list" and reverted to the rank of Private. He was, once again, appointed U/L/Cpl on 9 May 1941. Later that month he was admitted to Bodmin Emergency Hospital for 7 days for the treatment of tonsillitis. With effect from 25 January 1942, he was granted pay of appointment. As a paid Lance Corporal, he volunteered for flying duties as a Glider Pilot and was transferred to the Army Air Corps (AAC) and posted with the Glider Pilot Regiment for a period of 21 days, with view to transfer. On 27 August 1942, he gained another stripe and was appointed Acting Unpaid Corporal (A/U/Cpl) which was customary for trainee Glider Pilots and granted pay of appointment the same day. By the end of September 1942, he had obviously failed to meet requirements and relinquished the rank of Acting Corporal on suspension from flying. He reverted to the rank of Private. In December 1942 he was transferred and posted to 9th Battalion, the Royal Berkshire Regiment from 2nd Battalion Glider Pilot Regiment. During his time with the Royal Berkshires, he was transferred once more to 5th Battalion and also spent some time in hospital with influenza. He took part in the Normandy landings on D-Day (6 June 1944) when 5th Battalion, the Royal Berkshire Regiment, formed part of No. 8 Beach Group. Donald Noble was finally transferred to 4th Battalion, the Wiltshire Regiment, in September 1944, after No. 8 Beach Group had been abolished. His service record reports him killed in action on 4 October 1944 in Northwest Europe.

Summarising his record of service:

- 1938–1940: 2nd and 9th Battalion Royal Warwickshire Regiment (boy, orderly room clerk)
- 1940–1941: 70th Battalion (young soldiers) Royal Berkshire Regiment (Pte & L/Cpl)
- 1941–1942: 10th Battalion Royal Berkshire Regiment (Pte & L/Cpl)
- Oct 1942: Attached No. 1 wing the GPR in order to become a Glider Pilot (Cpl)
- Dec 1942: Transferred to 9th Battalion Royal Berkshire Regiment (Pte)
- April 1943: Transferred to 5th Battalion Royal Berkshire Regiment (L/Cpl)
- 5 June 1944: Embarked for Normandy – 5th Berkshire Regiment, No. 8 Beach Group – (L/Cpl)
- Sept 1944: Transferred to 4th Battalion, the Wiltshire Regiment (L/Cpl)

14.8 Genealogy

With no dentals available, the only means of identification left would be a DNA test, for which next of kin would be required. RIU's UK house genealogist Lt. Col (ret.) David Chilton in Salisbury (UK) was asked to help find the next of kin of L/Cpl Donald S. Noble. He found Donald's entry of birth at the General Register's Office (Photo 14.8).

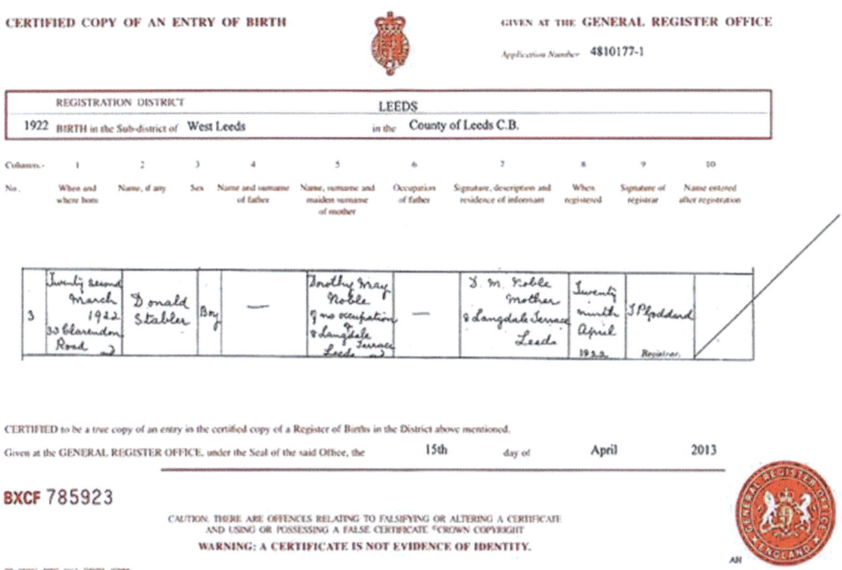

CERTIFIED COPY OF AN ENTRY OF BIRTH

GIVEN AT THE GENERAL REGISTER OFFICE

Application Number 4810177-1

REGISTRATION DISTRICT LEEDS

1922 BIRTH in the Sub-district of West Leeds in the County of Leeds C.B.

Columns:- No.	1 When and where born	2 Name, if any	3 Sex	4 Name and surname of father	5 Name, surname and maiden surname of mother	6 Occupation of father	7 Signature, description and residence of informant	8 When registered	9 Signature of registrar	10 Name entered after registration
3	Twenty second March 1922 30 Clarendon Road	Donald Stabler	Boy	—	Dorothy May Noble of no occupation of 8 Langdale Terrace Leeds	—	D. M. Noble mother 8 Langdale Terrace Leeds	Twenty ninth April 1922	J Plodland Registrar	

CERTIFIED to be a true copy of an entry in the certified copy of a Register of Births in the District above mentioned.

Given at the GENERAL REGISTER OFFICE, under the Seal of the said Office, the 15th day of April 2013

BXCF 785923

CAUTION: THERE ARE OFFENCES RELATING TO FALSIFYING OR ALTERING A CERTIFICATE AND USING OR POSSESSING A FALSE CERTIFICATE ©CROWN COPYRIGHT

WARNING: A CERTIFICATE IS NOT EVIDENCE OF IDENTITY.

Photo 14.8 Birth certificate for Donald S. Noble

Donald Stabler Noble was born 22 March 1922 in the parish of West Leeds, Yorkshire. His mother was Miss Dorothy May Noble, but his father was not registered. Possibly his surname was Stabler, considering Donald's second Christian name and Stabler being a common surname in the north of the United Kingdom. Dorothy May was a daughter of Harry Douglas and Miriam Noble. They had four other children. Three of Dorothy's siblings (Miriam Ethel, Douglas and Edna) never had any children. The fourth sibling, Gladys Lee, who died in 1983, did have a daughter by the name of Elizabeth Mary. Elizabeth Mary gave birth to a daughter, Elizabeth Lindsey, in 1947. She would both have been a suitable donor for a mitochondrial (Mt DNA) DNA test, since she shares the Noble Mt DNA with Donald. But both Elizabeth Mary and her daughter, Elizabeth Lindsey, turned out to be untraceable. According to some sources, the family emigrated to Australia in the 1950s. Several unsuccessful attempts were made to trace Elizabeth Lindsay in Australia and New Zealand. She was believed to have married a gentleman by the name of De Vries (of Dutch descent); however all attempts to contact the lady in question were left unreplied.

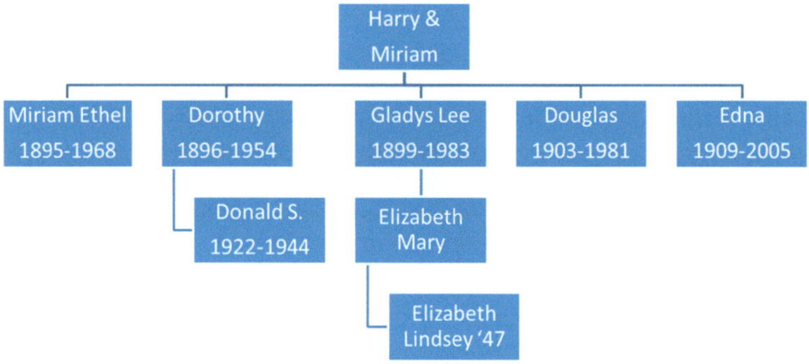

With no living DNA donors available, there was but one option left, which was to open Dorothy Noble's grave in Lawnswood Cemetery in Leeds and obtain a DNA sample from her remains for matching purposes. Ms. Dorothy May Noble had died on 9 December 1954 in a Leeds hospital. She was the owner of the grave, which she shared with her parents Harry and Miriam Noble. On the headstone she had also commemorated her missing son Donald with the following words: "In proud and loving memory of L/Cpl Donald Noble, missing in action at Arnhem Oct 4th 1944, aged 22 – Abide with me".

14.9 Exhumation in Leeds, UK

With a covering letter from the British Defence Attaché to the Netherlands, Captain Nigel Amphlett (Royal Navy), and the help from his Defence Team, requests for an exhumation were submitted to the Leeds City Council Bereavement Services, the Chancellor of the Ripon and Leeds Diocese and the Ministry of Justice, Burial and Cremation Policy Team London, all in order to obtain permission for the partial exhumation of Dorothy May's remains in order to obtain DNA samples for DNA testing purposes (grave S550 in the Lawnswood Cemetery in Leeds (UK). After paying a fee for a Faculty to the Diocese and providing a Risk Assessment to the Yorkshire Environmental Health Department, the licence to exhume was finally issued in March 2014. On 9 April 2014, the RIU team were able to collect DNA samples from the remains of Dorothy May Noble. The partial exhumation of the remains of Dorothy May Noble took place on 9 April 2014 in Lawnswood Cemetery in Leeds (UK). All dental elements turned out to have been extracted during life; therefore two bone samples (7 cm shafts of both humeri) were collected (Photos 14.9 and 14.10).

14.10 DNA Research

Of SR 006/13 a dental element, a sample of the left femur and a sample of the cranium were submitted for DNA testing. This was conducted by the Netherlands Forensic Institute (NFI) in The Hague, the Netherlands. From SR 006/13 an autosomal DNA profile was established of 7 loci and a Y-chromosomal profile of 21 loci. From Dorothy May's remains, an autosomal profile was established of 15 loci. Based on the autosomal DNA profiles, the likelihood of SR 006/13 and Dorothy May Noble being related was 100.000–1.000.000.000 times larger than the two being not related. This implied a strong enough match.

Summarising, the biological, medical and historical profiles were fully consistent. All information considered it was concluded that SR 006/13 was indeed Donald Stabler Noble, 1922–1944 (Photo 14.11).

14.11 Conclusions

The second set of skeletal remains, SR 007/13, were identified as those of Private Harold Lewis from Hertford (1925–1944). Both soldiers were reinterred with full military honours in Oosterbeek CWGC War Cemetery on 5 October 2016, by a burial party provided by 5th Battalion, the Rifles.

Additionally, the following outcomes were obtained:

Photo 14.9 Opening of the grave in Lawnswood Cemetery, Leeds. (Photo: RIU)

- Mt DNA analysis using 70-year-old biological material is quite possible, even despite apparently poor quality of bone and teeth samples.
- Trotter (1970) stature reconstruction proves (again) to be accurate. Only a 4 mm discrepancy between the calculated and AM living stature was found: SR, 170,4 cm; AM stature, 170,8 cm.
- In obtaining DNA samples for reference purposes in historical casework, one needs to dare think outside the box and be prepared to make the effort: perseverance is not a sprint, it's a marathon (Photo 14.12).

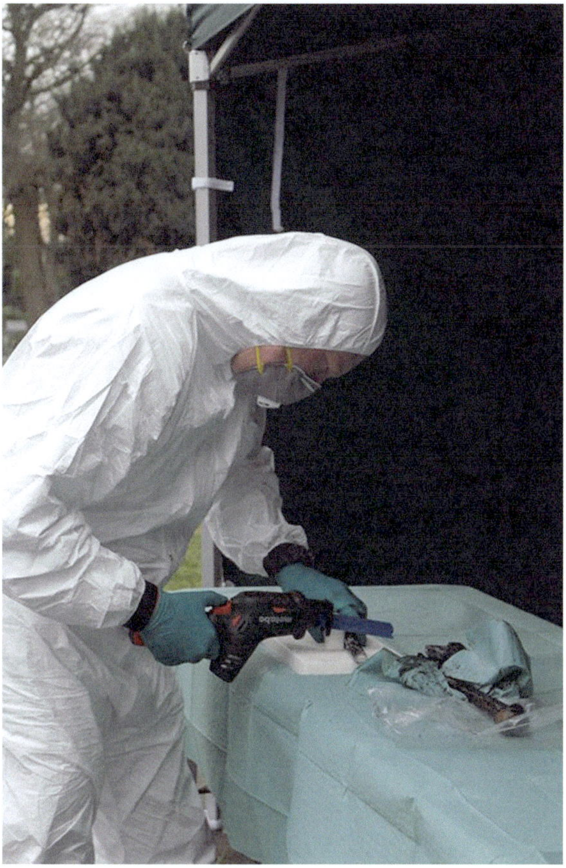

Photo 14.10 Taking bone samples from both humeri

Photo 14.11 Headstone of the Noble family grave. (Photo: RIU)

Photo 14.12 Reinterment ceremony of Donald Noble and Harold Lewis in Arnhem-Oosterbeek War Cemetery

Chapter 15
The Independent Commission for the Location of Victims' Remains in Ireland: A Case Study – The Abduction, Murder and Secret Burial of Victim B

Geoffrey Knupfer, Dennis Godfrey, and Jon Hill

Abstract Since the seventeenth century, when Britain laid claim to Ireland, relations between Ireland and Britain have been characterised by sometimes violent disputes over sovereignty. In the early part of the twentieth century, Ireland gained its political independence from the UK (of Great Britain and Ireland) with the exception of an area in the north-east of the island of Ireland – Northern Ireland, which remained part of the UK. The constitutional arrangement for Northern Ireland has been in place since then.

Keywords Northern Ireland troubles · Good friday agreement · Commission for the disappeared · ICLVR · Paramilitary murders · Clandestine burials · Peat bog burials

15.1 Introduction

For nearly three decades, from the late 1960s, Northern Ireland was engulfed in a violent conflict that saw over 3800 people killed and many thousands more injured, both physically and mentally. In the main the violence was perpetrated by heavily armed republican paramilitary groups, the largest being the Provisional Irish Republican Army (PIRA), which sought to end the link between Northern Ireland and the UK and so-called loyalist paramilitary groups, which claimed loyalty to the UK. Republican violence was primarily targeted against the police and the British Army, the latter having been deployed to support the police and to protect the economic infrastructure in Northern Ireland. Loyalist paramilitaries focussed their violent attention primarily against those they loosely identified as republican.

G. Knupfer (✉) · D. Godfrey · J. Hill
Independent Commission for the Location of Victims Remains (ICLVR), Dublin, Ireland
e-mail: geoff@knupferconsulting.com

P. M. Barone, W. J. M. Groen (eds.), *Multidisciplinary Approaches to Forensic Archaeology*, Soil Forensics, https://doi.org/10.1007/978-3-319-94397-8_15

Of all those killed during what became known as the 'Troubles', one group stands out from all the others. Between 1972 and 1985, 15 men and 1 woman were abducted, murdered and secretly buried. One man was a serving soldier. All were taken by republican paramilitaries, and they are known collectively as the 'Disappeared'.

Over the years, many efforts were made by the UK and Irish governments and often with the support of the administration in the USA and other politicians, to establish constructive negotiations between the various groups and factions in the search for peace and a political solution. Numerous attempts failed, but, in 1998, peace negotiations finally began in earnest. They culminated in the signing of the Good Friday (or Belfast) Agreement, a peace treaty which, despite various hiccups in the intervening period, has now lasted some 20 years.

In seeking a settlement in Northern Ireland, both the British and Irish Governments recognised that the Disappeared was a specific humanitarian issue that needed to be addressed, as did the Republican Movement, which agreed to work to resolve it. As an integral part of the Good Friday Agreement, various independent commissions were established by the governments to monitor progress and to resolve outstanding and contentious issues impacting upon the progress of peace. One such Commission was the Independent Commission for the Location of Victims' Remains (ICLVR), or 'The Commission for the Disappeared' as it is commonly referred to in Ireland.

15.2 The Independent Commission for the Location of Victims' Remains

The Commission's role and remit was established by the enactment of legislation in Ireland and the UK. In both pieces of statute, which largely mirror one another, disappeared victims are described as individuals who were abducted and murdered prior to 1st April 1998 (the date of the Good Friday Agreement) by or on behalf of a 'Proscribed Organisation' (paramilitaries). Surprisingly, no commencement date was set out in this legislation. Given Ireland's turbulent history, this could have presented untold problems. Fortunately, however, this hasn't proved to be the case. The Commission's role is entirely humanitarian. It is tasked with recovering victims and returning them to their loved ones. It does not have any criminal investigative role. Any information received can only be used to recover and repatriate victims, and such information cannot be disclosed to any other individual or organisation. Information received by the Commission is inadmissible in criminal proceedings. Some observers have likened this process to an amnesty whereas in fact it more closely resembles a form of strictly limited immunity.

The Independent Commission commenced its work in 1999 under the direction of two Commissioners, each government appointing one. Whilst the Commission's operations remain totally independent of governmental control, all its running costs are met by the two governments. In its early stages, the Commission was established as a largely reactive organisation, awaiting information to be passed to it. To some degree the structure proved reasonably effective in that two former paramilitary organisations declared knowledge of a total of ten cases and disclosed broad locations of their burial sites. Following those disclosures, three victims were recovered and repatriated. As the years passed, however, little further progress was made and, in 2006, following a review of its procedures, the Commission was transformed into a proactive organisation with its own investigative and search capabilities. Since that date several more victims have been recovered and repatriated. The Commission's initial list of disappeared victims totalled 16; however, upon closer scrutiny in 2006, it was found to contain some duplication, and the total number was reduced to 14. Then two further (and until that stage) completely unconnected cases came to light in 2010, and the total number was again increased to 16.[1] Thirteen victims have now been recovered and repatriated. Only three cases remain outstanding.

15.3 The Abduction of Victim B

Victim B, a male, was 23 years of age when he was abducted. At the time, he resided with his young wife in a flat in Belfast. He was not known to have been involved in any form of paramilitary activity. Just prior to his disappearance, he had worked in a local carpet factory but, when that business closed down, he obtained employment as a merchant seaman on a ship sailing between Northern Ireland and Rotterdam. He was due to commence this new job at some stage in April 1978.

On the morning of Saturday 8th April, B visited Belfast City Centre with his mother to do some shopping, leaving his pregnant wife at home alone. Shortly after his departure, there was a knock on the door. Upon answering it B's wife was confronted by nine men who restrained her, bundled her into the flat and asked for the whereabouts of her husband. When she told them he was out shopping, they injected her with some form of drug, restrained her and then lay in wait for B's return. When he did return a few hours later and entered the flat, they overpowered him manhandled him out of the flat and drove him away in a car telling his wife that he would be back later and warned her against reporting this to the police. He didn't return and was never seen alive again.

[1] As outlined, the Commission's list of victims currently totals 16. This is based wholly on the fact that the legal cut-off date (enshrined in both sets of legislation) for victims to be recorded as a 'disappeared victim of the troubles' is 1998, the date of the Good Friday Agreement. In fact, a further incident occurred in 2003 in which another male victim was abducted and murdered by Republican paramilitaries and his body secretly concealed. His remains were recovered in 2005. Whilst, in accordance with the Commission's legislation, his name does not appear on its list, in every other respect he too is one of the 'Disappeared', and his case is often included on other versions of the list of disappeared victims of the troubles.

15.4 The Searches for Victim B

When, in 1999, the Provisional IRA released its list of individuals it had abducted and killed, B's name appeared on that list. At that stage paramilitaries informed family members that B had been killed as 'An Agent Provocateur'. Subsequently, details of the believed location of his grave were passed to the Commission by intermediaries. The location was given as a peat bog in Oristown, Co. Meath, Ireland, approximately 110 km from B's home in Belfast (Fig. 15.1).

Upon initial inspection, this was found to be a vast area of bogland. The general location indicated is within a triangular piece of the bog bordered by roads on two sides and tracks on the third. At its widest point, it measures marginally over 1 km. The bog itself is divided into dozens of separate peat banks each individually owned and used for the harvesting of peat for fuel for open fires. Over the years, many drainage ditches have been dug, and these criss-cross the bog. It was suggested that B's grave was located on one of these peat banks in the general proximity of a drain. In this phase of the investigation, several banks were eliminated as being too close to the road, whilst others were highlighted as possible grave sites. It was stated that the victim had been walked over from the road to the pre-dug grave and shot in situ. The grave was then refilled and the surface reinstated to conceal the location. It seems the grave had seldom, if ever, been revisited or inspected by those responsible since the date of the event.

In 1999, prior to the Commission possessing its own search capability, information was passed to An Garda Siochana (The Irish Police Service) who, acting as agents of the Commission, undertook a physical search of various areas of the bog, based broadly on the above information and directions provided by intermediaries.

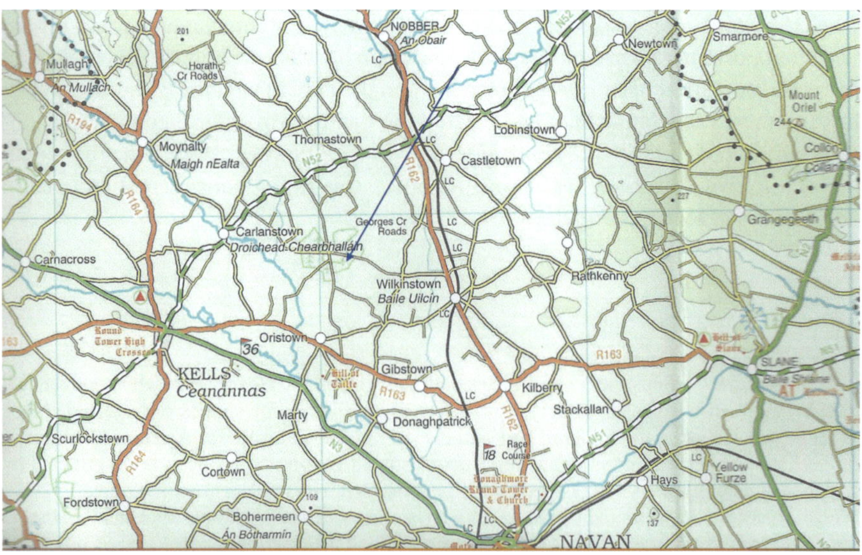

Fig. 15.1 1:150,000 map of general area of Oristown Bog, Co Meath

This search failed to locate the victim. In 2010, following further discussions with intermediaries, the search area was extended and mapped and a 20 × 20 m search grid superimposed on same. This process was followed by a cadaver dog search and geophysical survey of the site. A further physical search was undertaken by the Commission's in-house forensic archaeologists and contractors. This search, which lasted several months, also proved negative.

15.5 The Recovery of Victim B

In 2014, further and more refined information was received from an intermediary. It highlighted a reasonably specific area of the bog as the possible grave site. This area had not previously been searched. Following initial geophysical survey, site work commenced on the clearance and deepening of a number of existing drainage channels which run around the periphery of the peat banks. This was undertaken in an effort to facilitate the removal of water prior to the commencement of a comprehensive and structured search of the area indicated. This drain clearance was undertaken with mechanical excavators and the entire process supervised by forensic archaeologists, in accordance with standard Commission operating procedures. On the 1st October 2014, in the course of this clearance operation human remains were exposed in the side of a drainage channel running along the edge of bank number 6 (see Fig. 15.2). Whilst something of a surprise in the initial stages, it soon became

Fig. 15.2 An aerial image of Oristown peat bog looking south-east and showing the peat banks and drains

apparent that, over a period of many years, this drainage channel had probably been cleared and deepened and the banks either cleared or had otherwise eroded. What had been part of the surface of the bog at the time of the burial in 1978 had now become part of the bank of the drain. The scene was preserved and the State Pathologist summoned. Following recovery of the remains and post-mortem examination, it was established that the victim had died from a gunshot wound to the head. Clothing recovered with the body closely matched the description of the clothing worn by Victim B at the time of his abduction (Fig. 15.3 and Fig. 15.4).

When the Commission established its proactive capability in 2006, it also created a confidential familial DNA database. This was achieved by obtaining buccal swabs from all surviving close biological relatives of the missing victims and having these profiled. Following the above recovery, samples were taken from the remains and submitted for DNA profiling. The result was compared with samples previously provided by a sister and two brothers, and this produced a likelihood ratio of 550,000, indicating that the recovered remains were 550,000 times more likely to be those of a sibling of the victim's sister and brothers than those of another person. Taken in conjunction with the information provided as to the location of the remains, the clothing recovered with the body and the fact that no other family siblings were unaccounted for, the Coroner and the Inquest jury accepted this evidence as an unequivocal identification of Victim B.

In the course of its three physical searches of the bog, the Commission had excavated, searched and reinstated a total of 6.5 h of land. Following a religious service, Victim B was placed to rest in a family grave alongside his mother and father.

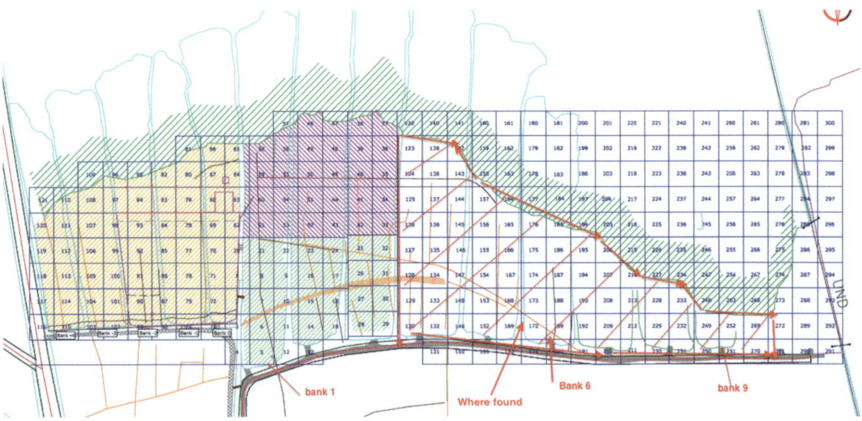

Fig. 15.3 Plan of search areas. The three stages of the 2010 search are shown in yellow green and purple. The intended search area for 2015 is shown in white with red border. Each grid measures 20 × 20 m

Fig. 15.4 The grave upon initial discovery (footwear and clothing can be seen lower right)

15.6 Discussion

B's family had engaged with the Commission from the outset, appreciating that a Commission recovery would limit the chances of a criminal prosecution. They realised that if they didn't accede to the legal terms and conditions set out in the Commission's legislation, their chances of ever recovering their loved one would be remote. Whilst the incident is recorded as a murder in both jurisdictions, an active criminal investigation has not been pursued.

In recovering Victim B, the Commission team worked entirely within its legal structures. Other than the post-mortem examination (which was undertaken to establish the cause of death and as part of the identification process), evidence that would normally form part of a criminal investigation was not recorded, seized or retained. None of the specific information received by the Commission from intermediaries was passed on to the police or any other agency or individual. No one has been arrested or prosecuted for the murder of Victim B.

Due to the fact that some 36 years had elapsed between the abduction and recovery of Victim B, a number of problems were encountered in the course of the search process. It was clear, for example, that the terrain of the search area had changed somewhat in the intervening period – vegetation mentioned by some participants to intermediaries had clearly been removed at some stage, possibly as part of the mechanisation of peat harvesting from the site. This difficulty has been encountered in a number of other Commission cases where terrain has changed dramatically over the interval between burial and search or recovery. In one case, for example, a now mature conifer forest had been planted on or adjacent to the suggested burial site. Contrary to early fears, the Commission has never been in receipt of misinformation from former paramilitary participants as to the location of graves. We are satisfied that there has never been any intention to deceive. On the contrary, some former participants have gone to significant lengths to assist this process. It has been suggested that this may be some form of cathartic process on the part of some individuals.

An associated problem is the memory of those who participated in these events and who now provide information to the Commission via intermediaries. Given the timescales involved in these cases, some are less sure of detail, and others clearly are unable to recall exact locations. We suspect also that there might have occurred instances of 'recollection by committee' where perhaps the loudest voice rather that the most accurate recollection prevailed. In other cases, those directly involved in these events have died. In summary these processes are invariably protracted and challenging and seldom accurate or specific. Despite thorough initial surveys, the decision to excavate often results in protracted physical searches of extensive tracts of land.

One of the other complicating factors regularly experienced by Commission staff is managing the expectations of the family of the missing victim. The moment staff start to undertake physical survey and search, there is an assumption the Commission is in possession of precise information and that it will only be a matter of days or perhaps a couple of weeks away before their missing loved one is found and repatriated. This seldom proves to be the case. In the search for Victim B, for example, many months of physical search was undertaken in three separate stages over a period of 15 years.

In every Commission case where physical search is considered, contemporary records, local knowledge, personal accounts and aerial imagery are closely scrutinised. Initial survey work is always undertaken, be that ground walking by forensic archaeologists, canine search by cadaver dogs or geophysics or a combination of all three. Whilst canine search has a good track record in locating recently buried human remains, this has not proved the case so far as the Commission is concerned as the most recent abduction and murder on its books occurred over 30 years ago, and some of its outstanding cases occurred well over 40 years ago. It is speculated that these poor results are due to remains of that age being in a relatively stable state and where the active decay process has substantially diminished.

Geophysical survey and search has also been undertaken to some degree at every search site. Once again, results have not proved encouraging. Whilst other organisations appear to have achieved good results, so far as multiple burials are concerned, it seems that a single burial where the remains are skeletonised and have been in situ in excess of 30 years will always prove something of a challenge. In this particular case, the site was surveyed with ground penetrating radar (GPR) prior to physical search commencing. Unfortunately, however, since the remains were recovered in the side of a drainage ditch, rather than on the flat, this survey did not extend to the actual grave site. The Commission is often assured by experts that geophysics is a well tried and tested technology that should produce results, even in the somewhat testing circumstances of the Irish 'Disappeared'. Sadly, this has not proved to be the case in our experience.

15.7 Conclusion

There is a strong body of opinion to suggest that the Commission's legal structures do facilitate the recovery of victims, but at a significant cost, since families know that, in participating, it is unlikely that anyone will ever be brought to justice for the abduction and murder of their loved one. To that degree, the legal framework within which the Commission operates constitutes a conspicuous compromise. Nevertheless, experience has shown that it is a compromise to which all the (Commission's) families are prepared to sign up. Time and again, following a successful recovery, family representatives have publically thanked those who provided the information which led to the return of their loved one. This would seem to suggest that this process of 'truth and reconciliation' appears to work. It might also explain why representatives from a number of former conflict zones around the world are closely monitoring the work of the Independent Commission for the Location of Victims' Remains.

Part III
Forensic Archaeology and Antiquity Crimes

Chapter 16
Not Just Body and Decomposition: Forensic Archaeology Preventing Antiquity Crimes

Pier Matteo Barone

Abstract The term forensic archaeology also identifies a sector that uses archaeological, scientific, and legal disciplines to study and analyse the strategies and modus operandi of criminal groups which endanger and undermine cultural heritage through theft, looting, and national and international trafficking.

A forensic archaeologist serves as an expert or consultant to the court, investigators, law enforcement, lawyers, and other figures, institutions, local authorities, or private individuals and must demonstrate the origin of archaeological finds and art subject to illegal excavations or illicit trafficking and draw up appraisals and assessments on damage to cultural heritage and seized archaeological finds and artwork.

Keywords Forensic archaeology · Forensic investigation · Cultural heritage protection · Antiquities crime · Illicit trafficking · Archaeological heritage looting

When archaeological methods are involved in forensic investigations, they are usually used to search for missing bodies, which are generally buried underground, and/or to analyse discovered bodies. Recently, these techniques have become relevant for other purposes as well. Archaeological approaches have become as important in numerous forensic cases 'without a body' as they have been in 'normal' criminal cases, particularly in Italy and in the Middle East. If the general guidelines followed during the crime scene investigation (CSI) are the same, the differences in the approach are relevant.

A forensic archaeologist's activities in this sector are quite varied and include demonstrating the origin of archaeological finds and art subject to illegal excavations or illicit trafficking and drawing up appraisals and assessments on damage to cultural heritage and seized archaeological finds and artwork. Therefore forensic archaeology also indicates the professional activities carried out in this field,

P. M. Barone (✉)
American University of Rome, Rome, Italy

Geoscienze Forensi Italia® – Forensic Geoscience Italy, Rome, Italy
e-mail: p.barone@aur.edu

P. M. Barone, W. J. M. Groen (eds.), *Multidisciplinary Approaches to Forensic Archaeology*, Soil Forensics, https://doi.org/10.1007/978-3-319-94397-8_16

whether in or out of court, by archaeologists, art historians, and other professionals working to protect cultural heritage as experts or consultants to the court, investigators, law enforcement, lawyers, and other figures, institutions, local authorities, or individuals.

If the normal study of archaeological materials is aimed at reconstructing their historical context and events, in legal archaeology the evidence regarding a find's history must also include information about present events: the identification of clandestine excavation areas and illicit traffic, the damages suffered by their contexts, and the possibility of reconstructing them.

At times, some countries refer to this field as judicial archaeology instead of forensic archaeology, creating confusion, improper uses, and disparate meanings.

Mainly in Italy (but not exclusively), the aforementioned term (judicial archaeology) is used in journalism to indicate dated legal cases, defining them as 'the archaeology of justice'; this meaning was also taken up by Judge Ferdinando Imposimato in one of his books: after having recently escaped the failed attack in Addaura, Giovanni Falcone confronted Imposimato, declaring that he was considered a fool in the salons of Palermo for wasting his time getting mixed up in old stories and 'instigating judicial archaeology.'[1]

It is basically for this reason that forensic archaeology is also referred to when there is a lack of a classic crime scene (CSI) at an international level: its purpose is to fight crimes against cultural heritage by consulting various professional figures who can together carry out investigations and work on the prevention and repression of vandalism and crimes against cultural heritage.

In fact, the above term is used to identify all those criminal associations, mainly of a mafia type, that manage the international illicit trafficking of cultural assets, from clandestine excavations to the black market. It is well known that art is third in the ranking in mafia trafficking, preceded only by drugs and weapons; it is also the fourth most profitable market for international crime.

In fact, investigations, seizures, and recoveries of antiquities by the judiciary system and the police are certainly commonplace, with the relative civil and criminal judicial disputes arising from the same. The result is a new professional field of archaeology and other disciplines relating to cultural heritage: more and more often archaeologists, art historians, and cultural heritage professionals in general are called to intervene in prevention, investigation, and repression activities for crimes against cultural heritage and in judicial proceedings related to the seizure or recovery of archaeological finds and works of art circulating in the clandestine market. In comparison with other disciplines used in the criminological and judicial field, we can define the use of the methodological approach of forensic archaeology as a professional field that brings the skills of archaeologists and other cultural heritage professionals into the judicial field to establish the origin of archaeological finds and art subject to illegal excavations or illicit trafficking and draw up appraisals and assessments on seized archaeological finds and artwork or the entity of damages to cultural heritage.

[1] F. Imposimato, 2013. *The Republic of Unpunished Massacres*, Newton Compton.

Working in the judiciary field as technicians and professionals at the service of law enforcement, judges, lawyers, institutions, local authorities, or private individuals to carry out the prevention, investigation, and repression of crimes and acts of vandalism against cultural heritage requires, in addition to excellent preparation in the field, knowledge, abilities, and specific skills that the normal academic path in the field of archaeology and cultural heritage today does not provide, as indeed already occurs for many other areas of professional archaeology, thus following the risk of facing the working world with an education that is technically inadequate for carrying out very delicate roles that also involve significant legal responsibilities. In fact, if archaeologist's studies are usually aimed at reconstructing historical contexts and events, in forensic archaeology it is essential to know how to analyse archaeological finds and works of art to derive useful information for drafting expertise on cultural assets for sale, surveys of archaeological finds, and seized works of art; for identifying their origin from illegal excavations, theft, or other illegal activities; for helping investigators reconstruct the channels of illicit traffic; and for making estimates and economic evaluations of assets subject to litigation and the damage caused to cultural heritage by criminal activities and acts of vandalism.

16.1 The Italian Situation

From a legislative point of view on the subject, Italy is at the forefront at an international level. The cultural heritage and landscape code (also known as the Urbani Code from the name of the then Minister of Cultural Heritage and Activities, Giuliano Urbani) is an organic body of provisions relating to the cultural heritage and landscape assets of the Italian Republic which was issued with Italian Legislative Decree No. 42 of January 22, 2004.

The code identifies the need to preserve Italian cultural heritage. It considers cultural assets as all tangible and intangible things of artistic, historical, archaeological, or ethno-anthropological interest; this also includes architectural assets, collections of cultural institutions (such as museums, archives, and libraries), naturalistic assets (such as mineralogical, petrographic, palaeontological, and botanical goods), and historical scientific assets, geographical maps, and photographic material (photography, negatives, and audio-visual (cinematographic film)). Intangible and landscape assets are also considered as being of cultural interest.

Not just laws but also operational departments: in Italy we have a single police force which works exclusively in this sector, more specifically the Carabinieri for the Protection of Cultural Heritage (TPC). It works under the Minister for Cultural Heritage and Activities as a directly collaborating office and conducts judicial police investigations, working against all the violations of cultural heritage law carried out by individuals or criminal organizations, in particular clandestine excavations at archaeological sites; theft and receipt of artwork and related illegal trade; damage to monuments and archaeological areas; illegal exportation of cultural goods; falsifications of antiques and other works of painting, graphics, and sculpture; recycling operations carried out through the reinvestment of proceeds from the illicit trafficking of cultural assets; and crimes against the landscape.

The TPC Carabinieri serve as a centre for information and analysis in the specific field in favour of all the departments of the Carabinieri and the other police forces and periodically perform monitoring and control of terrestrial and marine archaeological sites and UNESCO World Heritage Sites with the support of the Carabinieri Aircraft Group, the Carabinieri Horseback Group, as well as terrestrial and marine departments, all in constant collaboration with the Minister for Cultural Heritage and Activities and the relevant superintendencies.

They carry out constant control and detection activities at exhibitions, fairs, and auctions both in Italy and abroad, as well as with dealers, antique dealers, restorers, and art dealers, also maintaining close contact with the ecclesiastical authorities for the protection of ecclesiastical cultural heritage.

The TPC Carabinieri also operate abroad through Interpol and maintain relationships with other international police forces, participating in conferences, seminars, and research in areas of interest and collaborating with universities, foundations, and national and foreign research centres to conduct studies and develop operationally relevant projects. Organizations such as UNESCO, ICCROM, ICOM, UNIDROIT, and ICOMOS constantly collaborate with each other and with this nucleus to develop training and awareness activities for the public and industry operators at an international level. Specifically, a memorandum of understanding was stipulated with ICCROM (International Centre for the Study of the Preservation and Restoration of Cultural Property) for the performance of collaboration, research, training, dissemination, and promotion activities concerning the protection and conservation of cultural heritage.

Given their extensive experience, the TPC Carabinieri lastly carry out specialist training activities for foreign state ministry officials, police, and customs officers in the framework of specific cultural cooperation agreements at a government level or at the request of international organizations and intervene in crisis areas in the context of international peace missions to protect the concerned countries' cultural heritage, assisting the officials of the Minister for Cultural Heritage and Activities and the superintendents in activities relating to surveys and the securing and recovery of cultural heritage in areas of Italy which have been affected by serious disasters.

The aforementioned Code has been amended on numerous occasions up to the present day. Just before the end of 2017, the Council of Ministers of the Italian Republic reformed, on the proposal of the Minister of Cultural Heritage and Activities and Tourism and the Minister of Justice, the legislation on crimes against cultural heritage, with the declared objective of increasing penalties and introducing more effective instruments for combating the illicit trafficking of cultural goods. The new aspects of the Code include the introduction of crimes relating to illicit trafficking, illicit detention, and theft of cultural assets, as well as a very significant novelty on the prevention front relating to crimes for the unjustified possession of metal detectors.

This is a further sign that institutions in Italy have recently begun to focus more on criminal activities linked to the illicit trafficking of art and antiquities, which with their annual turnover of between 6 and 8 billion dollars constitute the most profitable organized crime business worldwide after drugs and weapons. Moreover, Italy is one of the most affected countries: it has an average of three ascertained crimes a day, and the subtraction of cultural assets from the community is estimated at around 150 million euros a year. Fortunately, despite often inadequate legislation, Italy is also among the first in the world for its ability to combat the illicit trafficking of art and antiquities and their recovery.

16.2 Case Studies

Numerous forensic cases, in which archaeological approaches have been used, have become as important as the criminal cases, particularly in Italy. In this section, two examples will be discussed.

Both of them involve the Ministry of Cultural Heritage, which places legal restrictions on areas with high archaeological potential. Based on this, nobody, including the landowners, can remove soil for any purpose, from building new construction to planting. In these cases, private landowners can confirm the presence or absence of archaeologically relevant remains beneath their soils. The non-destructive methods and the archaeological evaluation are the best options to help a proper act to preserve and protect antiquities in an adequate way.

16.2.1 Southern Italy

In the first case, the private landowners of this area had the opportunity to confirm the presence of archaeologically relevant remains beneath their soils using ground-penetrating radar (GPR) and to ask the superintendent and the ministry to remove or reduce the legal restriction to a more well-defined area.

In particular, the archaeological find consists of a series of post holes relevant to huts, clay fragments from ceramic dish mixing, some botanical remains, and remains of decorative elements dating back to protohistory. The discovery has been widely documented and is almost entirely stored in the local museum, excluding the largest wall structures. In this regard, it was necessary to perform a careful analysis of the historical satellite images to determine how the effective portion of land had been excavated and how extensive the archaeological finding in relation to the cadastral excerpt present in the report of the superintendent was (Fig. 16.1).

The satellite photos of the area show that, from 1989 to 2007, the research was focused mainly on a certain part of the sandy plateau, confirming what was documented by the temporary expropriation of the superintendent. In particular, just

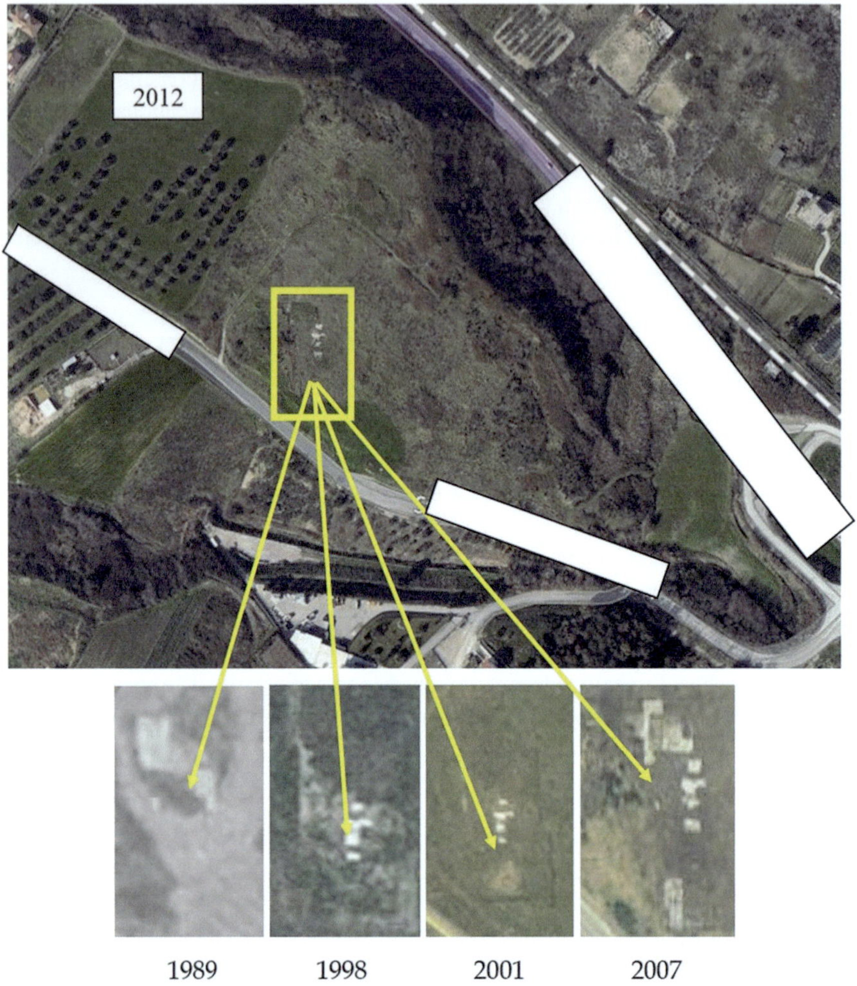

Fig. 16.1 The main figure shows the plateau with the archaeological restriction, in particular the archaeological site as it appeared in 2012 (in yellow). In previous years, the only area of the plateau that changed is the area affected by the temporary expropriation of the superintendent. In the bottom of the figure, it is easy to recognize the evolution of this area (and the archaeological excavation) from 1989 to 2007 (the year of the backfilling)

three cadastral particles were investigated by archaeology for a total of approximately 2 ha.

After 2007, when the temporary expropriation was finished, the excavations were backfilled, and the most relevant remains were collected (Fig. 16.2). In 2014, the superintendent together with the Ministry of Cultural Heritage decided to place more severe legal restrictions in the area. The restrictions do not allow building or cultivating or, generally, removing soil from the interested area.

Fig. 16.2 From the top, the final stages of the excavation of the archaeological site in 2007; in the middle and at the bottom, the area as it is today after the backfilling using white nonwoven fabric

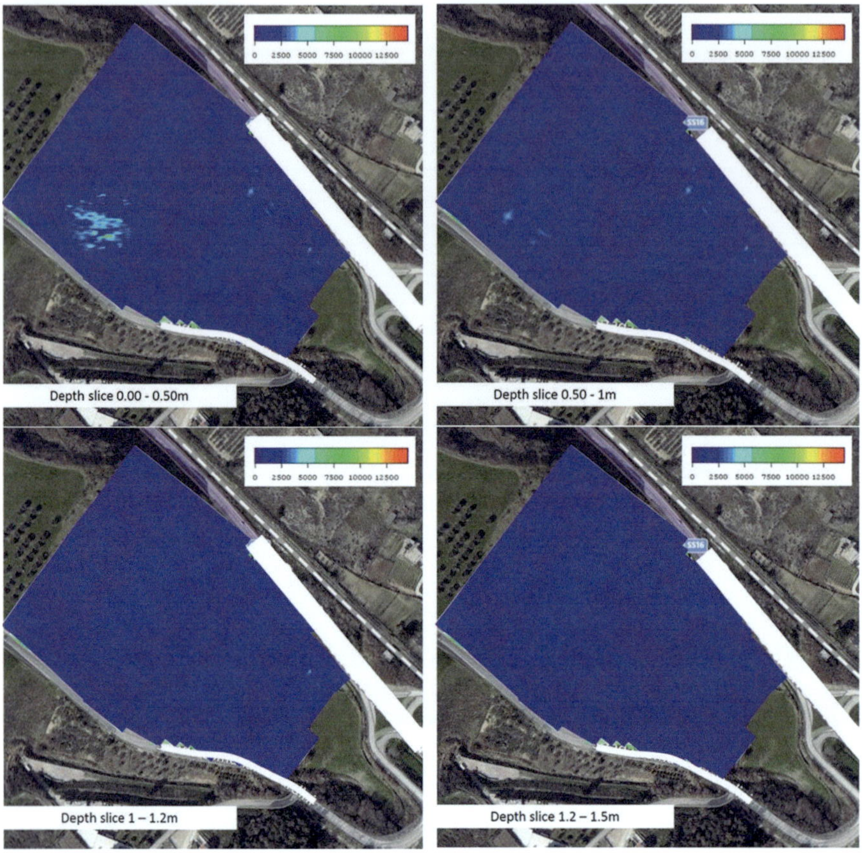

Fig. 16.3 In the figure, it is possible to note not only the dimensions of the plateau on which the Ministry of Cultural Heritage put the restrictions due to archaeological potential issues but also the relevant results of the forensic GPR investigations. The archaeological remains were not deeper than 1 m ($v = 0.12$ m/ns). Based on these results, the Ministry has reduced the restriction from 11 to 2 ha

The area of interest was clarified in 2007 as a result of the careful analysis of the satellite photographs, the field walking, and the GPR surveys to demonstrate how the areal dimensions of the superintendent's restriction was objectively too large.

The area affected by GPR prospections was approximately 11 ha and demonstrated how the use of this geophysical tool has allowed both high-resolution and detailed results. In fact, the GPR depth-slice results at different depths (up to 3 m from the surface) (Fig. 16.3) showed no particular anomalies. The only anomaly is the strong presence in the subsoil of the remains previously excavated and documented by the superintendent to a depth of 0.10–0.50 m.

The GPR investigations confirm the archaeological presence in only a limited area (2 ha) and not deeper than 1 m and highlight the complete absence of other remains beneath the soil.

In this first case, the GPR survey scientifically refuted the possibility of finding further archaeological remains in the investigated area and reset the level of archaeological potential in the area, giving the Ministry of Cultural Heritage and its superintendent a unique possibility to better define the restrictions.

16.2.2 Northern Italy

This case demonstrates how if on the one hand Italy is truly at the forefront of legislation and operations regarding the protection of archaeological heritage and on the other there is a lack of specific training in forensic archaeology; this results in the preparation of archaeological reports which are fallacious and deleterious to a correct forensic investigation.

This study initially envisaged carrying out a thorough analysis of all the documentation collected in relation to the case of an alleged illegal excavation in an area subject to archaeological restriction.

This case provided for a forensic inspection of the areas which were subjected to the alleged violation, also supported by GPS technology (Global Positioning System) in order to accurately map and locate the areas concerned. Furthermore, a small GIS (geographic information system) was created in order to visually clarify the geographical position of the violated areas.

This study showed that both the reports which had been drafted and the technical-archaeological surveys not only had evident signs of incongruity, imprecision, and at times fallacies but also demonstrated a lack of technical and scientific rigour, limiting themselves to simple and subjective assessments full of reasonable doubts.

Moreover, this report showed the presence of an archaeological restriction in the area in question but in an area that was only partially affected by the investigation in which there was neither an enclosure nor any public signage informing of this restriction. Furthermore, it shows how the discovery of the seized finds was fortuitous, and not necessarily carried out within the restricted area (Fig. 16.4).

16.3 Conclusions

Good archaeological expertise can lead to different conclusions with respect to initial impressions and the initial configuration of crimes. Although there may be some procedural gaps in a system for the protection of cultural property that is theoretically more than valid, and although these procedural flaws can be greatly diminished, if not eliminated, with the correct use of forensic archaeology in the field of cultural heritage protection and antiquity crime, it is imperative to firmly emphasize that the possession of an archaeological find is a serious offence both on a normative level and on an ethical and moral level, as it is always the result of clandestine

Fig. 16.4 The route (in blue) to take on foot to reach the area under investigation. The white circle and red circle always, respectively, represent the lack of a part of the walls and the location of the intervention. At the bottom, on the right, and on the left, some photos of the area are shown which demonstrate the total absence of signs and/or fences

excavation and therefore contributes to the destruction of the cultural sites and heritage of an entire nation, if not of humanity.

16.4 In Memory of

On July 26, 2017, a hero from another era was lost: a pioneer and protagonist in the fight against crimes against cultural heritage. This man was Roberto Conforti. He died at the age of 79, 42 of which he worked in the Carabinieri (from 1961 to 2002), dedicated to fighting crime and protecting cultural heritage, remaining loyal to the State and its institutions even when they turned against him.

He led the Carabinieri for the Protection of Cultural Heritage from 1991 until his retirement. His leadership brought the Carabinieri to become the world's leading police corps for the fight against crimes against cultural heritage. He and his 300 soldiers carried out hundreds of operations and recovered thousands of archaeological finds and works of art, including a number of impressive ones such as the Capitoline Triad, a number that is symbolically well-suited to his heroic and unequalled struggle against criminals of art.

After having left the Carabinieri, he continued his commitment to raising awareness for the protection of cultural heritage as President of SIPBC – the Italian Society for the Protection of Cultural Heritage.

In dedicating this chapter to his memory, the author would like to remember him with some of his brief but incisive considerations on Italian cultural heritage: "Italy is a collection of dense cultural contexts, with strong specific connotations and deep historical roots. From the Alps to Pantelleria, there is no corner lacking testimony of this thousand-year-old civilization, whose expressions are not just a legacy of antiquity, but play a significant role for the cultural and economic growth of the nation. The immense heritage is widespread in space and interacts with the land, both in its urban and landscape context. It is varied in relation to the typologies that characterize it. It is purely ours, as unlike in other countries, it is made up of assets produced in our own land. It is immanent to the geographical area it was created for. It is placed in such a way as to create a chronological cultural testimony. It is an expression of our identity".[2]

[2] R. Conforti, 1998. *Criminal law concerning the protection of cultural heritage*, in A. Balloni (ed.), Criminology and Security, Milan, pg. 95.

Chapter 17
Problems of Archaeological Heritage Protection: The Background of Crime Against Monuments and Works of Art

Maciej Trzciński

Abstract This article is devoted to selected issues in combating and preventing crime against archaeological heritage. The author presents basic legal provisions determining archaeological relics and defines the term "crime" in relation to them. Whilst describing basic activities of state institutions within the scope of combating this type of crime, the author initiates a discussion on insufficient staffing of police monument coordinators. The most controversial examples of prosecuting attorneys' errors are mentioned in the discussion on certain procedural problems, which have appeared in the course of conducting penal procedures known to the authors. The issue of participation of archaeologists and museum employees in the illegal trading of archaeological relics "in good faith" was also discussed. Moreover, the authors described the problem related to the absence of efficient solutions with respect to trade in archaeological relics from illegal excavation sites outside of Poland. The article ends with a summary of combating crime against archaeological relics and a conclusion that Poland is still missing the proper determination and, more importantly, proper evaluation of these types of crime and awareness of the importance of the problem. Negative effects of criminal activities are overlooked or marginalized; it should be remembered that such activities generate nonrenewable losses.

Keywords Cultural heritage · Criminalistics · Crime · Protection

The protection of cultural heritage is a complex issue concerning not only proper legal regulations but also an efficient state and local administration. It is also clear that the effectiveness of the cultural heritage protection system will be determined by appropriate funding.

Criminology and criminalistics have been efficiently and directly engaged in activities related to cultural heritage protection for years. This concerns not only actions related to the detection of offenders, proving their guilt and recovering stolen

---

M. Trzciński (✉)
Faculty of Law, Administration and Economics, Department of Forensic Sciences, University of Wrocław, Wrocław, Poland

P. M. Barone, W. J. M. Groen (eds.), *Multidisciplinary Approaches to Forensic Archaeology*, Soil Forensics, https://doi.org/10.1007/978-3-319-94397-8_17

objects, but also the protection of cultural heritage.[1] In an increasingly frequent manner, technical and forensic methods are used for marking and thereby protecting works of art. As we know, forensic tests are also used effectively in the process of confirming the authenticity of artefacts and works of art.

Cultural heritage, especially its material components in the form of antiques (artefacts and monuments), has been threatened for centuries in the same manner. Some of the dangers they face are grouped and are as follows:

The first group consists of hazards linked with the impact of natural elements (earthquakes, volcanic eruptions, acid rains and so on).

The second group is generated by humans. Among them, invariably across centuries, we find (1) wars, (2) crimes and offences committed in connection with the undertaking of various enterprises (e.g. construction, agricultural or forestry works) and (3) crimes as a result of planned and organized activities (theft, burglary, smuggling, fraud, handling stolen goods, robbery, vandalism and forgery) in situations when the subjects of the crime are antiques (artefacts and monuments) or works of art.

Speaking about crime against artefacts or monuments and works of art, it should be noted that, unfortunately, under Polish legal regulations, we do not have a uniform system of terms defining legal protection.

For example, the Constitution refers to national heritage and goods of culture. However, the main legal act regulating in detail antiques' protection, i.e. the Act from 23 July 2003 on antiques'[2] protection and care, refers to antiques and archaeological artefacts (and monuments). The Polish Penal Code,[3] on the other hand, uses the term "goods of particular importance" to refer to culture.

Thus, we have, under national law, the following terms: cultural heritage, national heritage, goods of culture, antiques, archaeological artefact and monuments as well as extralegal terms such as a work of art and old goods, which are not legally defined but, unfortunately, often used in everyday language, as they are synonymous with goods of culture or antiques.

It is obvious that under the application of law (especially criminal law), we must define precisely what is the subject of a crime. For example, a person who committed an antique forgery would be treated differently than a person who forged a picture of a living contemporary artist.

This issue appears to be extremely important, especially for prosecutors and judges engaged in such criminal proceedings. Unfortunately, under Polish conditions, frequent discord results due to a lack of precision concerning these terms.

Now, when examining some crimes whose targets are antiques and works of art, some attention should be paid to the following questions: What are the motives of the perpetrators? Why do people steal, falsify and smuggle antiques and works of art?

---

[1] M. Trzciński, O. Jakubowski (red.) Przestępczość przeciwko dziedzictwu kulturowemu. Diagnoza, zapobieganie, zwalczanie. Wrocław 2016.

[2] Ustawa z dnia 23 lipca 2003 r. o ochronie zabytków i opiece nad zabytkami (art.3 pgf.1)-**Dz. U. Nr 162 poz.1568**.

[3] Ustawa z dnia 6 czerwca 1997 r. Kodeks karny -Dz. U. Nr 88, poz..553.

Without a doubt, the basic motive of criminal activity is profit. Art is the currency of the world—this nineteenth-century concept explains everything. Skillful investment in antiques and works of art is a profitable business. In connection with this concept, on an increasingly frequent basis, the countries of Central and Eastern Europe are also dealing with the phenomenon of "money laundering" through the medium of buying works of art and antiques.

The money from arms, drugs and human trafficking is used in the art market (antiquarian market), and this is an international phenomenon.

It appears, however, that economic reasons are not the only reasons that motivate people to commit crimes. Quite a large group of offences targeting antiques and works of art have connections with the perpetrators' psychopathological disorders. Here, we are dealing with a morbid collecting passion—objects are stolen and then held in private collections.

An example of this phenomenon is Stephane Breitwieser, who, between 1991 and 2001, committed a total of 174 thefts (from museums, galleries and private collections), and stole 240 objects, whose value was estimated at 10 million euro. Breitwieser did not sell the stolen objects; he collected them for pleasure and kept them in his home.[4]

Sometimes antiques and works of art are destroyed by a person affected by religious mania, and this particularly concerns religious art. It may also happen during some kind of artistic event (as was the case in 2000 in one of Warsaw's galleries, where a visitor destroyed a work of installation art representing the pope crushed by a meteorite).

At this point, it is worth mentioning the phenomenon of art-napping, when a stolen object becomes the subject of blackmail and where a thief demands a ransom for its safe return and threatens to destroy the object if his or her demands are not met.

Acts of terror, whose victims were not people, but rather antiques, works of art or historical monuments, have also taken place.

Recently, there have been many acts of terror against cultural heritage in Iraq, Afghanistan and Syria (Palmyra). In 1993, a car bomb exploded in front of the Uffizi Gallery in Florence (Italy).

What do the crime statistics against antiques and works of art look like?

According to police statistics in Poland, which are disclosed annually, close to 1500 of these types of crimes were committed.[5] This places Poland, when compared with Interpol data on such crimes, at the forefront of these crime statistics.[6]

What is very worrying is that the data from the Ministry of Justice found the concerned people guilty and sentenced them. This data confirms observations that,

---

[4] V. Noce, Kolekcja egoisty, Warszawa 2008.

[5] In 2005 noticed -2247, in 2013–1819, in 2014–1295 crimes against foods of culture.

[6] About statistics, for example, O. Jakubowski, Criminal threats to the national heritage-an analysis of the events of 2014./in:/Santander Art. and Culture Law Review, Nr 1/2015, p. 263–274., O. Jakubowski, Criminal threats to the national heritage-an analysis of the events of 2015./in:/ Santander Art. and Culture Law Review, Nr 1/2016, p. 241–258.

first, the police encounter problems with the detection of offenders (especially those who target archaeological sites). Second, there are evidence issues faced by the prosecutors and judges. In my opinion, criminal cases are wrongly discontinued too often. This is largely because judges and prosecutors still consider such cases as holding minor importance.

What are some specific crime against antiques and works of art?

I would like to recall some examples:

A classic one was a theft, which occurred in 1911, from the Louvre. A museum employee, Vincenzo Perugia, stole the Mona Lisa by Leonardo da Vinci and transported it from France to Italy, where he tried to sell it. The perpetrator left his fingerprints at the crime scene, being unaware that the French police were already using fingerprint identification. Arrested in 1913, the perpetrator testified that the motive for his actions was patriotic.

Another example of a spectacular crime was the second theft of Edvard Munch's painting "The Scream" in 2000 from the National Gallery in Oslo.

In 2003, a gold mannerist sculptural masterpiece, Benvenuto Cellini's "Salt Cellar" (Saliera) was stolen from the Kunsthistorisches Museum in Vienna. The perpetrator executed the theft successfully, even though the alarm system was turned on.

In the national museum in Poland, we had an interesting case in 2000, when a perpetrator carried out the theft by exchanging an original Claude Monet painting for a counterfeit that he had made.

We also came to know of some worrying examples from Poland, where in the last few years, there have been several thefts from Auschwitz-Birkenau, the Nazi extermination camp. In 2009, the notorious theft of the iconic sign, which hangs over the gate with the inscription Arbeit macht frei, took place. The man who funded the theft turned out to be a Swedish collector.

This example shows that there are no locations which are immune from these types of crimes. It appears that ethical and moral principles are not barriers. Art is a commodity.

Archaeological sites are another target of criminal activities.

The distinct nature of these places is mainly on account of their unique locations. The majority of archaeological sites are located underground or underwater.

Whilst antiques held in museums, galleries and private collections are catalogued, archaeological artefacts found in old cemeteries or settlements are not itemized in any way, nor do they have accurate documentation. It is difficult to provide effective, physical protection measures for all underground and underwater archaeological sites.

The specific nature and distinctiveness of this class of antiques is, therefore, bound with their location and to the fact that they have not been categorically identified. Unfortunately, these features are exploited by criminals.

Another very important feature of archaeological artefacts is the cultural context in which they are located. What matters most is where a given artefact was located and under what circumstances it was discovered. An archaeological artefact "cut off" from its original context becomes, at the most, an attractive object, whose history will remain forever unknown.

Therefore, attention must be paid to the fact that in the case of archaeological artefacts, the most important aspect is not their material or artistic value, but rather their scientific and historical one.

The problem of archaeological heritage protection in Poland has been a very serious one for a considerable number of years. The main threat is created by illegal excavations.

It is said that Poland is struck by the "Indiana Jones" syndrome. About 100,000 people participate in treasure hunting as a hobby.

It should be noted that exploring artefacts requires an appropriate licence that is issued by a Regional Office of a Heritage Conservator. Undertaking such a search without a licence is an offence. In a year, about 50–60 of this particular type of licence is issued in Poland. Estimating the exact number of illegally conducted searches has been very difficult to date. Unfortunately, Polish law has not introduced any regulation with regard to the purchase of metal detectors or their use. Therefore, proving someone has used a metal detector for locating artefacts is an arduous task.

There is no doubt that trade on the Internet has led to an escalation of this kind of crime. Online transactions are international, fast and seemingly anonymous. Thus, on the Polish antiques market, there are archaeological finds possibly available from Ukraine, Bulgaria, Hungary and even perhaps Romania. Poland is an EU member state and belongs to the Schengen Agreement. Border controls, therefore, have been substantially liberalized, which, of course, favour smuggling activities (Fig. 17.1).

Fig. 17.1 Looted archaeological artefacts revealed by the police during the search

Unfortunately, cases concerning crimes against archaeological sites and artefacts are still very few (about ten per year). To detain a perpetrator for such a crime and prove his guilt still poses a difficulty for the criminal justice system.

These problems are of an objective nature but also, unfortunately, concern actions incompetently carried out by the prosecution that tends to marginalize the social harm of these crimes.

Our current knowledge about the perpetrators is rudimentary. Often, they are men aged between 20 and 60 years, with secondary and sometimes even higher education and no criminal record. The perpetrators have not only been unemployed people but also, for example, a student of archaeology, a policeman and even a priest!

All the nabbed perpetrators have traded the illegally obtained artefacts in order to obtain financial benefits. The majority of transactions have been conducted over the Internet, and some of them have occurred in the so-called flea markets.

It has to be emphasized at this point that, under Polish law, the sole owner of discovered or found archaeological artefacts is the State Treasury (the State). The status of ownership defined in this way, in fact, excludes archaeological artefacts from trade.

One of the main problems related to the prevention of crimes against archaeological sites and artefacts is the still insufficient knowledge prevalent among police, prosecutors and judges, who do not possess practical experience in this area and are sometimes not able to properly interpret the legislation that is in force.

The issue of cooperation between professional archaeologists and conservator offices with amateur explorers is also unregulated.

Equally alarming is the situation when amateur explorers run illegal excavations on old battlefields, where soldiers' bodies are exhumed. In these cases, two offences may be committed: (1) illegal possession of firearms and ammunition and (2) the desecration of human remains (Figs. 17.2 and 17.3).

So, how to effectively prevent crimes against archaeological sites and artefacts?

One of the basic issues that requires to be addressed is the need to increase the professionalization of the police, who in this respect need to work with conservation authorities, border guards, customs offices and museums, which in many cases undertake the function of identification and valuation of artefacts.

Since 2007, a special group dealing with crime against cultural heritage has worked within the remit of the Polish police force. It is difficult to say if these actions are fully professional and sufficient.

A separate problem pertains to the actions of prosecutors and courts. It was as late as 2011, before regular training began for prosecutors and judges in this regard.

The number of cases improperly discontinued by prosecutors is alarming and certainly does not contribute to the effective prevention of crimes against antiques. These problems arise from ignorance and disregard for such matters. Apart from the legal regulations, which even EU member states define differently as the status of the ownership of archaeological artefacts, an important issue is society's participation in the protection of cultural heritage. Unfortunately, among the so-called post-socialist countries, which include both Poland and Romania, the level of awareness

Fig. 17.2 Looted archaeological artefacts revealed by the police during the search

Fig. 17.3 Damages caused by plundering excavations at the Bronze Age burial ground in Lower Silesia

of the legal and cultural issues in relation to the protection of cultural heritage (and especially the archaeological one) is still insufficient.

From 1 January 2018, illegal exploration will no longer be treated in Poland as a minor offence but as a crime. This change in criminal law should strengthen the protection of archaeological monuments. However, the most serious problem is the low level of legal culture. Poland has not yet managed to build a conscious civil society that protects its cultural heritage.

Archaeological heritage protection is a problem that goes beyond the obligations of one state or nation. Archaeological heritage is the foundation of the cultural identity of Europeans. Its losses due to destruction and theft are not only irreplaceable but also nonrenewable. Therefore, we need to protect our heritage jointly.[7]

---

[7] More about the specificity of crime against archaeological monuments; M. Trzciński, Przestępczość przeciwko zabytkom archeologicznym. Problematyka prawno-kryminalistyczna, Warszawa 2010.

Chapter 18
Preventing Art and Antiquities Crimes Using Forensic Geology

Rosa Maria Di Maggio

Abstract Geological materials, such as gems, precious stones, precious metals, rare fossils, and archeological objects, are extremely attractive to organized crime because of their high economic value, small size, and non-traceability. Because of the traceability of bank transactions at international levels, criminal organizations are using precious means of exchange for traffic in arms and drugs, for money laundering, and for illegal trade. In addition to these crimes, precious geological materials are often falsified, forged, and imitated by methods that have reached such advanced techniques as to produce specimens almost perfectly identical to their natural counterparts, to the point that it is not easy to distinguish the differences.

This paper focuses on the main characteristics of valuable geological materials (gems, precious metals, and rare fossils), which are also of archeological interest, their falsification, and the techniques of forensic geology used for their study.

Keywords Geological materials · Gems · Fossils · Precious metal · Illegal trade · Forgery

18.1 Introduction

Precious geological materials are all those natural materials, such as precious stone and metals, gems, and rare fossils, that thanks to their rarity and beauty have reached a great economic value.

The characteristics of these materials have made them extremely attractive to organized crime due to their high monetary value, small size, and non-traceability. In fact, criminal organizations are using precious means of exchange for traffic in arms and drugs and for crimes that require a monetary transaction, with the aim of avoiding the traceability of bank transactions at national and international levels. Gems, as well as jewels, precious metals, and works of art, are used for money

R. M. Di Maggio (✉)
Geoscienze Forensi Italia® – Forensic Geoscience Italy, Rome, Italy
e-mail: dimaggio@geologiaforense.com

P. M. Barone, W. J. M. Groen (eds.), *Multidisciplinary Approaches to Forensic Archaeology*, Soil Forensics, https://doi.org/10.1007/978-3-319-94397-8_18

laundering by criminal associations as an alternative form of investment for illegal proceeds. This aspect has dramatically increased illegal trading and smuggling, which involves countries such as China, Russia, Afghanistan, and many tax havens. For example, over approximately 4 years (2008–2012), 1.5 tons of gold worth more than 26 million euros were smuggled into Italy (Capuano 2010; Di Maggio et al. 2013). In addition to these crimes, there are also those relating to the fraudulent falsification and imitation of precious geological materials.

18.2 Historical Overview of the Falsification of Precious Geological Materials

Already in ancient times, it was customary to falsify gems; the earliest evidence of changing the appearance of gemological material dates back to the Minoan era. During the ancient Roman period, Latin writers Pliny the Elder and Fedrus recount how some artisans were reported because they faked silver objects in the names of two of the most famous Grecian artists Praxiteles and Myron (Di Maggio and Barone 2017; Giacosa 2017).

One of the most famous episodes of fraud by falsification was reported by the Latin writer Vitruvius. He relates that Archimedes started to study hydrostatics because he was asked by his cousin King Hiero the Second to determine whether his crown was made of pure gold or an alloy with silver (Di Maggio and Barone 2017).

It seems that Archimedes, pondering how best to solve the king's problem, went to the public baths for his daily bath. Archimedes noticed that the more his body sank into the water, the more water ran out over the sides of the tub. He realized that he had found the solution to Hiero's problem; Archimedes had discovered the method for measuring the volume of an irregularly shaped object. Archimedes knew that gold was denser than silver, so a piece of gold weighing a certain amount would be smaller than a piece of silver weighing the same. Thus, the total volume of a gold plus silver crown would be greater than the volume of a pure gold crown. Archimedes found that the crown displaced more water than a lump of gold of equal weight. Thus, he concluded that the crown was not pure gold.

From the ancient Roman Empire until the Middle Ages, the art of falsification of precious objects developed increasingly, and it also became interested in objects of art and religious items, such as artifact and relics. In particular, the interest in Christian artifacts, both economic and religious, developed in parallel with the dissemination of Christianity in the Mediterranean basin, and although in the first centuries after Christ these objects consisted of relatively poor materials, they were considered of great value from the religious and liturgical point of view (Giacosa 2017). The allocation of this intrinsic value generated the wide development of trade in forged or falsified precious religious objects. Until the end of the Middle Ages, the falsification of precious objects had never been counteracted.

Only during the fourteenth century, in an effort to prevent and counteract the falsification of silver objects, did hallmarks on precious objects start to be used

Fig. 18.1 Some of the most important hallmarks used in the Kingdom of Naples

(Fig. 18.1). Successively, the master silversmiths started to create their own hallmarks to prevent the selling of fake silver objects in their names (Giacosa 2017).

With respect to coins, since ancient times when money substituted for the barter system, many false gold and silver coins have been produced and issued into circulation. Regardless of the collection of old coins, falsification had already started in the Renaissance (with the birth of numismatic collecting); to complete coin collections with "missing pieces," fakes were artfully created (Giacosa 2017).

The first synthetic gems were produced in the mid-nineteenth century, and in the early years of the twentieth century, synthetic copies reached an acceptable quality.

Over the centuries, with the new knowledge of materials and the development of chemical and physical techniques, artificial alternatives based on lower-value materials have evolved as replacements for precious metals and gems.

18.3 Gems and Precious Stones

Gems are special minerals and, much more rarely, rocks that thanks to their beauty, hardness, and rarity have an extremely high monetary value. In general, a precious stone is defined as a mineral from which a gem can be extracted; the cut and finished gem itself is then considered as a mineral often used in jewelry. There are also gems of organic origin such as amber, coral, pearls, and ivory.

Since Neolithic times, mankind has vested particular interest in precious stones, using them not only as valuable ornaments but also as symbols evoking magical and spiritual virtues (Di Maggio et al. 2013). Unfortunately, gems and precious stones have not attracted people only for their beauty; because of their rarity and high economic value, the practice has evolved of resorting to various kinds of falsification.

18.3.1 Methods of Falsification

Falsification of gems occurs mainly by three methods: exaltation, synthesis, and replacement (Di Maggio and Barone 2017).

The exaltation of gems is a practice that allows for the embellishment of precious stones, enhancing the features already present or hiding defects that are commercially underappreciated to artificially transform a poor-quality gem into a more prized specimen. Exaltation has been already in use for several centuries, and it is mostly performed to improve or change the color of the gems. Color is a characteristic that enables the identification of a gemstone and the estimation of its value, although gems that belong to the same mineralogical species may have completely different colors. However, color is one of the most easily forged features. Today, there are many treatments for enhancing the color, using techniques of irradiation or heat treatment (Fig. 18.2).

Other types of exaltation involve procedures for hiding clearly visible fractures or cracks. In fact, gems have distinctive signs, which may be superficial (called surface signs) or internal (known as inclusions), and are formed during the mineral crystallization process. The purity of a gemstone relates to its degree of internal clarity and to the nature, number, arrangement pattern, and size of the solid and liquid inclusions. Usually, the greater the purity of a gem, the greater is its value, especially for diamonds.

In general, the various kinds of exaltation are easily recognizable with low-power microscopy, even though the techniques are in continuous evolution.

Synthesis consists of artificially reproducing gems, with specimens having the same chemical, physical, and optical properties as the corresponding natural gems. There are different synthesis processes; however, the most common one consists of melting a material with a composition equal to that of the imitated mineral and adding metal oxides to obtain the desired coloring. The first synthetic specimens were already being produced in the middle of the nineteenth century. The synthetic stones are quite recognizable under the microscope as they exhibit typical curvilinear striae of growth and sometimes gas bubbles (Fig. 18.3). In recent times, however, the methods of synthesis have achieved advanced techniques to obtain products almost perfectly identical to their natural corresponding minerals, to the point that the differences are not easily distinguished. In these cases, identifying the products of synthesis often requires the use of very advanced and expensive instrumentation.

Fig. 18.2 Sapphire exalted by heat treatment with the diffusion of foreign chemical elements; the color is not naturally homogeneous and is concentrated at the edges of the gems. Courtesy of the Gemological Institute CISGEM (Centro Informazioni e Servizi Gemmologici), www.cisgem.com

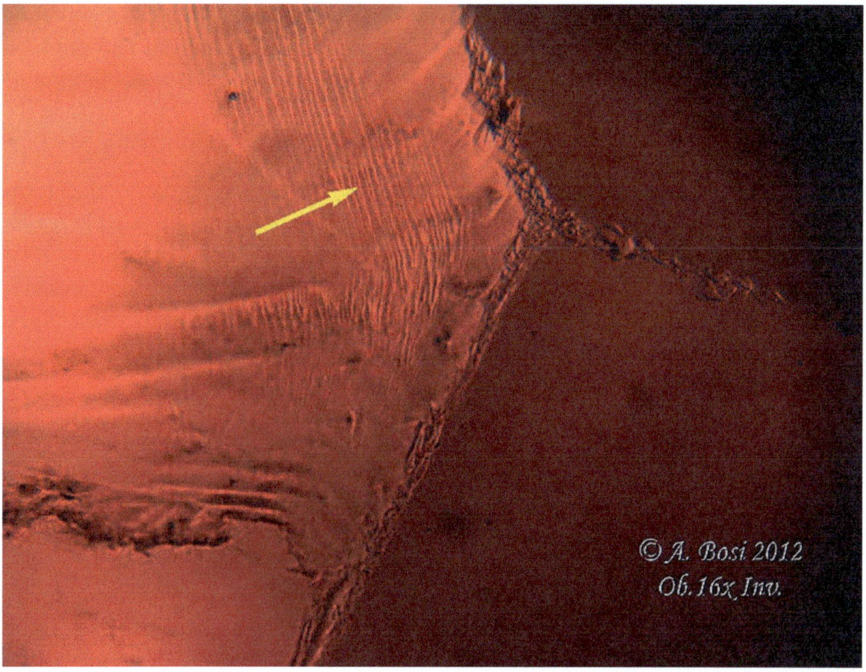

Fig. 18.3 Synthetic ruby at a magnification of 16×; curvilinear striae of growth are clearly recognizable. (Courtesy of Dr. Andrea Bosi)

Another counterfeiting method consists of replacing the most precious gems with relatively common natural stones of lesser value or with artificial materials, such as glass paste, ceramics, or plastic resins. A technique that allows saving or replacing precious material with low-quality material is that of the doublet, which consists of creating a gem made up of two parts; the upper one is normally a precious stone that is affixed to a lower part of lesser value. Sometimes, doublets of very precious stones are made exclusively with less precious materials. The aim of this forgery is to increase the carats of the gem. The carat is the unit of measurement for precious stones and is equivalent to one-fifth of a gram, thus 200 mg (Rolandi and Cavagna 1996; Giarin 2012). In general, other characteristics being equal, the greater the weight of the gem, the greater is its rarity, and, therefore, the higher is its value per carat.

18.3.2 Analytical Techniques

The analytical methods used to study the value of gemstones consist of the measurement of the refractive index, specific gravity, hardness, and caliber and the examination of all significant characteristics, such as color, clarity, brightness, cut, and distinctive signs (Rolandi and Cavagna 1996; Giarin 2012). In cases of imitation with synthetic products of the latest generation, advanced chemical and physical analyses are carried out on the gems to study their detailed features not detectable with superficial analysis or low-power magnification. These analyses envisage the use of spectroscopic but not destructive techniques, such as the following:

- Ultraviolet-visible-near infrared (UV-Vis-NIR) spectrophotometry; this technique registers the absorption phenomena of light radiation in the visible region of the electromagnetic spectrum (350–700 nm), in the near ultraviolet (200–350 nm), and in the near infrared (700–1100 nm). It is used mainly for the analysis of diamonds and the detection of their treatments; the determination of the geographical origins of emeralds, rubies, and sapphires; and the detection of some treatments on sapphires.
- Raman spectroscopy; this method analyzes molecular vibrations and/or rotations through the detection of the frequencies at which molecules diffuse radiation when they are subjected to laser radiation. It is used to identify heat treatments on diamonds and synthetic diamonds and various kinds of treatments on pearls, corals, emeralds, jades, and turquoises.
- Energy-dispersive X-ray fluorescence (EDXRF); this technique analyzes the elemental composition of a sample through the study of its X-ray fluorescence radiation. It is used mainly to establish the geographical origins of emeralds, rubies, and sapphires, detect the treatments of corundums and diamonds, and perform the analysis of pearls.
- Fourier-transform infrared spectroscopy (FT-IR); this method registers the vibrational phenomena of molecules when they are subjected to infrared radiation. It is mainly used to recognize resins and oils in emerald treatments and the treatments of opals, rubies, diamonds, turquoises, and jades and to analyze ambers.
- Photoluminescence spectroscopy (PL); this technique registers and detects the emission of photons from a material that is excited by electromagnetic radiation. It is used to detect the products of synthesis.

18.4 Precious Metals

The metals are elements distinguished by high thermal and electrical conductivity as well as by certain physical properties such as hardness and fusibility. Depending on their chemical stability, they are differentiated into precious metals and base metals. Precious metals are particularly resistant from the chemical point of view and possess exceptionally important physical characteristics, for example, ductility, which is the ability to withstand plastic deformation without rupture. In the commercial field, the well-known precious metals are gold, silver, platinum, and palladium.

Pure precious metals are too soft for the production of jewelry and objects; for this reason, they are alloyed with other metals that improve resistance and workability

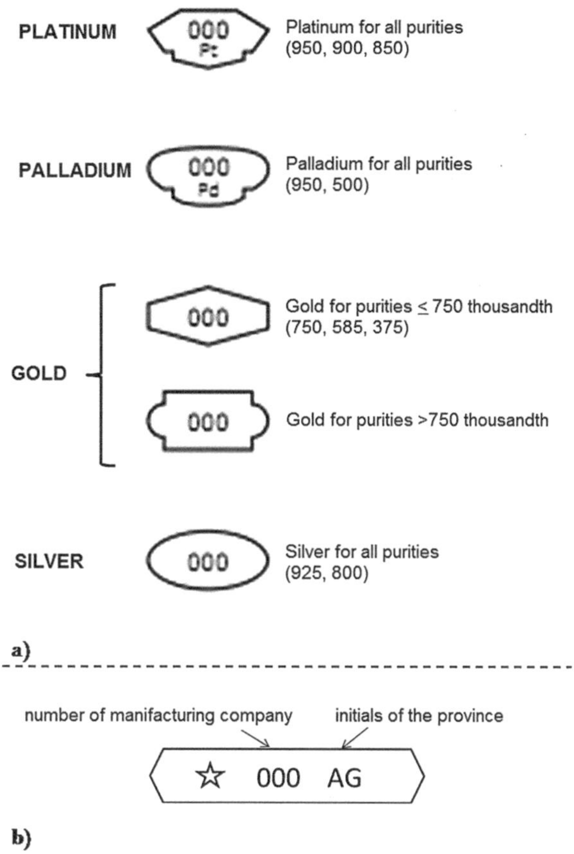

Fig. 18.4 Scheme of purities of precious metals and polygon shapes (**a**); identification mark (**b**)

and may also change the color depending on aesthetic choices; for example, the different shades of gold result from the different percentages of other metals that are added to pure gold.

Since a superficial examination cannot verify the precious metal content in an alloy, Italian legislation prescribes that objects made with precious metals and their alloys should be compulsorily inscribed with at least two signs that constitute the so-called hallmark: the indication of the purity and the identification mark.

The purity is expressed in thousandths and highlights the relationship between the masses of the base metal alloy and the pure precious metal (Di Maggio et al. 2013; Capuano 2010). For example, an 800 alloy contains 800 parts of pure precious metal and 200 parts of other metals. The purity number is inscribed in a polygon, the shape of which indicates the pure metal in the alloy (Fig. 18.4a).

The identification mark is represented by a polygon containing a five-pointed star, a unique number assigned to the manufacturing company by the chamber of commerce and the initials of the province where the company resides (Fig. 18.4b).

The hallmark allows the manufacturer to be traced and therefore held legally responsible for the purity compliance declared for the item.

18.4.1 Methods of Falsification

Generally, the falsification of precious metals occurs by substitution with less precious metals and subsequent plating with the metal to be imitated, as well as by employment of different alloy proportions compared to the specification of purity. Falsifications also involve the identification mark and the purity mark. Generally, the false hallmarks report incorrect purity marks, incomplete or missing indications of all the requested information for the precious object, numbers related to nonexistent manufacturing companies, etc. (Fig. 18.5).

Pure gold ingots are the subject of numerous fakes, as they constitute the standard of interbank trading. Years ago, the process used to falsify gold bars was to laminate with gold an equivalent volume of steel or lead. Steel and lead, however, have specific gravities less than that of gold; thus, the bars forged with this method weigh 60% less than the genuine version, making them easy to identify as fake. In the past decade, several falsified gold bars were discovered that were forged by the coating of tungsten blocks with a thin plate of pure gold. The choice of tungsten proved very advantageous because tungsten has a monetary value far less than gold but a similar specific gravity, so it is virtually impossible to distinguish a gold ingot from a tungsten forgery only by the weight/volume ratio. This forgery technique has provided an idea to those counterfeiters who had produced tungsten ingots using different methods.

Another type of falsification involves the forgery of ancient coins, which is particularly widespread all over the world. While the old fakes (e.g., the ones created during the Renaissance) have at present become documents of that period and therefore subjects of major interest, the modern falsifications are dangerous imitations that can turn into objects of fraud in sales if presented as genuine. Currently, the

Fig. 18.5 False engraving of the hallmark on a silver object; the purity mark is not a perfect ellipse, and the numbers are distorted although the surface is flat and relatively easy to engrave; the identification mark is not clear, and it does not allow identification of the company legally responsible for the declared purity

phenomenon has taken a more criminal connotation, and the falsification quality has improved to the point of requiring an expert to discern the true from the false one. Furthermore, within numismatic collections, the phenomenon of alteration of original coins has developed to improve their quality and to make them more desirable on the antiquities market. These alterations include the closing of holes; the facing of the coin funds; the addition of an artificial patina, which means the concretions formed on the surfaces of ancient objects as a function of the aging; the reconstruction of the hair of a portrait worn for the burin; and the alterations of the legends.

18.4.2 Analytical Techniques

Analytical methods for detecting precious metals are mainly the study of the hallmarks (the indication of the purity and the identification mark), which envisages the use of the stereoscopic microscope and/or the magnification lens and the analysis of the alloy and the plating. The last two methods involve different analysis procedures that range from quite straightforward and not invasive tests, such as the analysis of density or the so-called touchstone test, to spectroscopic techniques, used mainly to detect the exact chemical compositions of the alloys, the platings, and the cores. These methods include energy-dispersive X-ray fluorescence (EDXRF), scanning electron microscope and energy-dispersive spectroscopy (SEM-EDS), and inductively coupled plasma mass spectrometry (ICP-MS), which is particularly appropriate for white gold objects containing nickel or palladium.

The analysis of ancient coins, due to their high archeological value, involves nondestructive examination of the samples and their patinas. Generally, dating is performed on the concretions through ^{14}C, observation with UV light, and nondestructive spectroscopic techniques, such as XRF, SEM-EDS, and FT-IR.

18.5 Fossils

Many specimens of fossils have considerable commercial value due to their rarity, conservation status, and antiquity. Parallel to the great market demand for these specimens, their manipulation and/or falsification has become increasingly common and has improved over time, reaching such high levels that in a general examination, even the experts are misled.

Parallel to the falsification of fossils, in recent years, there has developed a contraband trade in fossils involving in particular dinosaur skeletons or parts of them, tusks and teeth of mammoths, ammonites, and fishes. The clandestine nature of the smuggling of precious fossils makes it impossible to calculate the dollar amount of world trade, but reliable industry expert estimates suggest that it is approximately tens of millions of dollars a year (Fig. 18.6) (Simons 2010).

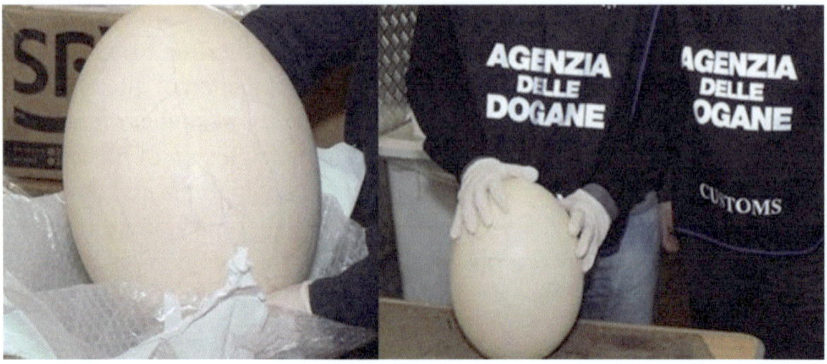

Fig. 18.6 Smuggled egg of *Aepyornis maximus*, an extinct species of giant bird that lived in Madagascar from the Pleistocene to the seventeenth century, on its way to the United States and seized by Italian customs

In Italy, any activity related to paleontology is bound by the laws of the Ministry for Heritage and Cultural Activities that govern paleontological research and activities and prohibit collection by anyone who does not provide serious and adequate guarantees.

18.5.1 Methods of Falsification

False fossils include both pieces reproduced more or less faithfully with various materials and pieces that have undergone a partial reconstruction or a touch-up, the addition of material details unrelated to the original fossil, or even recomposition in positions different from the original.

The counterfeiting of fossils aims to improve their appearance to tempt a buyer to purchase; for this reason, all fossils with commercial value are often subject to manipulation. The main forgery methods include the following:

- Reconstruction of missing parts; this forgery is performed when otherwise precious fossils have missing, ruined, or broken parts.
- Positioning of the fossil on a piece of matrix; this change occurs when fossils are found as loose materials in sediments.
- Assembly of pieces belonging to various individuals; generally, this forgery reproduces a specimen that is not in the same position as the original setting.
- Artificial coloring.
- Total reproduction.

The last method is usually carried out by means of a mold that can be produced using various materials; generally, appropriately colored resin is used. Resin is also used to make partial manipulations of the fossil. For example, during recovery operations, some types of fossil tend to detach from the matrix, as occurs, for example, with shark teeth; in these cases, the tendency is to reglue with resin the pieces that broke away, thus creating a falsification of the original situation.

18.5.2 Analytical Techniques

The methods used for the study of the false fossils are paleontological observations and chemical analysis, which include:

- The identification and classification of the specimen
- The taphonomy, which is the study of processes occurring from the death of the organism to its fossilization
- The association of species, which is useful in the case of finding several specimens for understanding if it is a natural or an artificial assemblage
- Chemical and spectroscopic analyses, which aim to determine the nature of the materials from which the false specimens are composed
- Dating analysis using radioactive isotopes including carbon 14

18.6 Final Remarks

Although forensic geology is a useful tool for analyzing precious materials and works of art, there are some critical issues that involve analytical procedures and their management. The falsification of gems, precious stones, and fossils by means of latest-generation products of synthesis and artificial materials has reached such advanced techniques as to produce specimens almost identical to their natural counterparts, to the point that it is not easy to distinguish their differences. In such cases, the experts in geology, gemology, or paleontology must use very advanced and expensive instrumentation with the aim of providing probative evidence of falsification and a definitive identification of the synthesis.

Furthermore, as far as precious metals are concerned, the critical issue of the analysis is that the objects subjected to examination often require technical procedures that change the integrity of the samples. Indeed, forged objects hide the evidence of falsification beneath plating, which is removed in performing the analysis. This procedure modifies the state of the samples under investigation and requires appropriate applications of legislative tools.

Bibliography

Abel RL, Parfitt S, Ashton N, Lewis SG, Scott B, Stringer C (2011) Digital preservation and dissemination of ancient lithic technology with modern micro-CT. Comput Graph 35(4):878–884

Abramov A, Veselovskaya E, Dolgov A, Engovatova AV, Mednikova MB, Nikitin S, Safarov A (2015) In: Groen WJM, Maquez-Grant N, Janaway RC (eds) Forensic archaeology in the Russian federation/forensic archaeology. A global perspective. Wiley Blackwell, Chichester, pp 139–149

Adams BJ, Byrd JE (2008) Recovery, analysis, and identification of commingled human remains. Humana Press, New York

Adams BJ, Byrd JE (eds) (2014) Commingled human remains: methods in recovery, analysis, and identification. Academic, Oxford

Ainsworth PB (2001) Offender profiling and crime analysis. Willan Publishing, Uffculme

Alberti B, Jones AM, Pollard J (2013) Archaeology after interpretation: returning materials to archaeological theory. Left Coast Press, Walnut Creek

Alfaro-Lucas JM et al (2017) Bone-eating Osedax worms (Annelida: Siboglinidae) regulate biodiversity of deepsea whale-fall communities. In: Deep Sea research part II: topical studies in oceanography. Available at: http://linkinghub.elsevier.com/retrieve/pii/S0967064517301297 Accessed 26 July 2017

Allard TT, Sitchon ML, Sawatzky R, Hoppa RD (2005) Use of hand-held laser scanning and 3D printing for creation of a museum exhibit. In: Mudge M, Ryan N, Scopigno R (eds) The 6th international symposium on virtual reality, archaeology and cultural heritage, pp 97–101

Allison PA, Briggs DEG (eds) (1991) Taphonomy: releasing the data locked in the fossil record. Plenum Press, New York

Anderson CC (2009) GIS spatial analysis of multiple scenes in criminal homicides, MA thesis, University of South Florida, Tampa

Anderson GS, Bell LS (2014) Deep coastal marine taphonomy: investigation into carcass decomposition in the saanich inlet, British Columbia using a baited camera P. Archambault, ed. PLoS One 9(10):e110710. Available at: http://dx.plos.org/10.1371/journal.pone.0110710 Accessed 26 Oct 2016

Anderson G, Bell L (2016) Impact of marine submergence and season on faunal colonization and decomposition of pig carcasses in the Salish Sea. PLoS One 11(3):e0149107. Available at: http://journals.plos.org/plosone/article?id=10.1371/journal.pone.0149107 Accessed 21 May 2016

Anderson GS, Bell LS (2017) Comparison of faunal scavenging of submerged carrion in two seasons at a depth of 170 m, in the strait of Georgia, British Columbia. Insects 8(1):33. Available at: http://www.ncbi.nlm.nih.gov/pubmed/28335403 Accessed 26 March 2017

P. M. Barone, W. J. M. Groen (eds.), *Multidisciplinary Approaches to Forensic Archaeology*, Soil Forensics, https://doi.org/10.1007/978-3-319-94397-8

Anderson GS, Hobischak N (2004) Decomposition of carrion in the marine environment in British Columbia, Canada. Int J Legal Med 118(4):206–209. Available at: http://link.springer.com/article/10.1007/s00414-004-0447-2 Accessed 20 May 2016

Andrews P (1995) Experiments in taphonomy. J Archaeol Sci 22:147–153

Appleby JEP (2016) Archaeothanatology in the English-speaking world: the belated spread and potential application of a methodologically rigorous approach to mortuary analysis. In: What bones tell us/El Que Ens Eqpliquen Els Ossos. SERP/University of Barcelona, pp 13–24

Aronson HS, Zellmer AJ, Goffredi SK (2016) The specific and exclusive microbiome of the deep-sea bone-eating snail, *Rubyspira osteovora*. FEMS Microbiol Ecol 49(3). Available at: https://academic.oup.com/femsec/article-lookup/doi/10.1093/femsec/fiw250). Accessed 26 July 2017

Ayers LE (2010) Differential decomposition in terrestrial, freshwater, and saltwater environments: a pilot study. Available at: https://digital.library.txstate.edu/handle/10877/4078. Accessed 22 May 2016

Baccino E (2011) L'Anthropologie forensique (AF) ou anthropologie médicolégale, est-elle un nouveau métier ? Rev Fr Méd Légale Rev Méd Légale 2:1–3

Baier W, Rando C (2016) Developing the use of structure-from-motion in mass grave documentation. Forensic Sci Int 261:19–25

Bang P (2014) Dobbeltdrabet i Tusindårsskoven. In: Nordisk kriminalreportage, pp 73–86

Barker P (1981) Tecniche dello scavo archeologico. Longanesi, Milano

Barker C, Alicehajic E, Naranjo Santana J (2017) Post-mortem differential preservation and its utility in interpreting forensic and archaeological mass burials. In: Shotsmans E, Márquez-Grant N, Forbes S (eds) Taphonomy of human remains: forensic analysis of the dead and the depositional environment. Wiley-Blackwell, Chichester, pp 251–276

Barone PM (2012) Archaeology, geophysics and forensic: united we stand, divided we fall. In: Proceedings of 31st Conference of the National Group of Solid Earth Geophysics (GNGTS 2012), Potenza 20–22 November. ISBN: 978–88–902101-3-6

Barone PM (2016) Understanding buried anomalies: a practical guide to GPR. Lambert Academic Publishers, Saarbrucken

Barone PM (2017a) Forensic archaeology. In: Di Maggio RM, Barone PM (eds) Geoscientists at crime scenes. A companion to forensic geoscience, soil forensics series. Springer, The Hague, pp 191–214. https://doi.org/10.1007/978-3-319-58048-7_9

Barone PM (2017b) Forensic geophysics. In: Di Maggio RM, Barone PM (eds) Geoscientists at crime scenes. A companion to forensic geoscience, soil forensics series. Springer, The Hague, pp 175–190. https://doi.org/10.1007/978-3-319-58048-7_8

Barone PM (2017c) Quality, expertise, and ethics in forensic geoscience. In: Di Maggio RM, Barone PM (eds) Geoscientists at crime scenes. A companion to forensic geoscience, soil forensics series. Springer, pp 215–224. https://doi.org/10.1007/978-3-319-58048-7_10

Barone PM, Di Maggio RM (2016) L'approccio alla scena del crimine tramite la geofisica forense ed i cani da cadavere, Il Penalista, Giuffré Editore. ISSN 2464-9635

Barone PM, Di Maggio RM (2017) Introduction to forensic geoscience. In: Di Maggio RM, Barone PM (eds) Geoscientists at crime scenes. A companion to forensic geoscience, soil forensics series. Springer, The Hague, pp 1–10. https://doi.org/10.1007/978-3-319-58048-7_1

Barone PM, Ferrara C (2017) The past beneath the present: GPR as a scientific investigation for archaeology and cultural heritage preservation. Int J Conserv Sci 8(4):581–588. ISSN: 2067-533X

Barone PM, Ferrara C, Pettinelli E, Annan AP, Fazzari A, Redman D (2012) Forensic geophysics: how GPR could help police investigations. In: Proceedings of the 18th European meeting of environmental and engineering geophysics of the near surface geoscience division of EAGE, Paris, 3–5 September 2012. https://doi.org/10.3997/2214-4609.20143307

Barone PM, Di Maggio R, Ferrara C (2015a) Forensic geo-archaeology in Italy: materials for a standardisation. Int J Archaeol 3(1–1):45–56. https://doi.org/10.11648/j.ija.s.2015030101.16

Barone PM, Di Maggio R, Ferrara C (2015b) Not necessarily buried bodies: forensic GPR investigations from criminal to civil justice. In: Proceedings of the 8th International Workshop on

Advanced Ground Penetrating Radar (IWAGPR), IEEE, Florence 2015, pp 1–4. https://doi.org/10.1109/IWAGPR.2015.7292681
Barone PM, Di Maggio RM, Ferrara C (2016a) Forensic geoscience during CS investigations and courtroom trials without a murder. J Forensic Sci Criminol 4(3):302. Annex Publishers . ISSN: 2348-9804
Barone PM, Swanger KJ, Stanley-Price N, Thursfield A (2016b) Finding graves in a cemetery: preliminary forensic GPR investigations in the Non-Catholic Cemetery in Rome (Italy). Measurement 80(February):53–57. https://doi.org/10.1016/j.measurement.2015.11.023
Barrios M, Wolff M (2011) Initial study of arthropods succession and pig carrion decomposition in two freshwater ecosystems in the Colombian Andes. Forensic Sci Int 212:164–172. Available at: http://www.sciencedirect.com/science/article/pii/S0379073811002763 Accessed 20 May 2016
Bartol CR, Bartol AM (2012) Criminal & behavioral profiling. Theory, research and practice. Sage, Thousand Oaks
Bartosiewicz L (2008) Taphonomy and palaeopathology in archaeozoology. Geobios 41:69–77
Bass WM (1997) Outdoor decomposition rates in Tennessee. In: Haglund WD, Sorg MH (eds) Forensic taphonomy: the postmortem fate of human remains. CRC Press, Boca Raton, pp 181–186
Beherensmeyer AK, Kidwell SM, Gastaldo RA (2000) Taphonomy and paleobiology. Palaeobiology 26(4):103–147
Behrensmeyer AK (1978) Taphonomic and ecologic information from bone weathering. Paleobiology Q J Paleontol Soc 4:150–162
Behrensmeyer AK (1985) Taphonomy's contributions to palaeobiology. Palaeobiology 11:105–119
Behrensmeyer AK, Kidwell SM (1985) Taphonomy's contribution to paleobiology. Paleobiology 11:105–119
Bell LS (2011) Histotaphonomy. In: Crowder C, Stout SD (eds) Bone histology: an anthropological perspective. Taylor & Francis, Boca Raton, pp 241–252 Available at: https://books.google.fr/books?hl=en&lr=&id=nS7P8-2UpUEC&oi=fnd&pg=PP1&dq=lynne+bell+Crowder+and+Stout&ots=g-lkNMEwh&sig=dndT_lhMp8Mjs7jW7dkVFUdX03g#v=onepage&q=lynne bell&f=false
Bell LS, Anderson GS (2016) The skeletal Histo-taphonomy of deep coastal marine submersion and exposure. In: American academy of forensic sciences, p 194 Available at: https://www.aafs.org/wpcontent/uploads/2016Proceedings.pdf
Bell L, Skinner M, Jones S (1996) The speed of post mortem change to the human skeleton and its taphonomic significance. Forensic Sci Int 82(2):129–140. Available at: http://www.sciencedirect.com/science/article/pii/0379073896019846 Accessed 20 May 2016
Bello SM, Parfitt SA, Stringer C (2009) Quantitative micromorphological analyses of cut marks produced by ancient and modern handaxes. J Archaeol Sci 36(9):1869–1880
Bello SM, De Groote I, Delbarre G (2013) Application of 3-dimensional microscopy and micro-CT scanning to the analysis of Magdalenian portable art on bone and antler. J Archaeol Sci 40(5):2464–2476
Benbow ME et al (2015) The potential of high-throughput metagenomic sequencing of aquatic bacterial communities to estimate the postmortem submersion interval. J Forensic Sci 60(6):1500–1510
Bennett R (2011) Archaeological remote sensing: visualisation and analysis of grass-dominated environments using airborne laser scanning and digital spectra data. PHD thesis
Bentley RA, Maschner HDG, Chippindale C (2009) Handbook of archaeological theories. AltaMira Press, Plymouth
Berger I, Redeker HT, Lawrance BN, Tague JT, Terretta M (eds) (2015) African asylum at a crossroads. Activism, expert testimony, and refugee rights. Ohio University Press, Athens
Binford LR (1977) General introduction. In: Binford LR (ed) For theory building in archaeology: essays on faunal remains, aquatic resources, spatial analysis, and systematic modellin. Academic Press, London, pp 1–10

Binford LR (1981) Bones: ancient man and modern myths. Academic Press, New York
Bintliff J (2004) A companion to archaeology. Wiley-Blackwell, Malden
Black SM, Aggrawal A, Payne-James J (eds) (2010a) Age estimation in the living: the practitioner's guide. Wiley-Blackwell, Chichester
Black S, Walker G, Hackman L, Brooks C (2010b) Disaster victim identification: the practitioner's guide. Wiley-Blackwell, Chichester
Böhler W, Marbs A (2004) 3D scanning and photogrammetry. In: Proceedings of the 12th international conference on geoinformatics. Gavle University Press, Gavle, pp 291–298
Borgys D, Kasen I, Achard V, Perneel C (2012) Hyperspectral anomaly detection: comparative evaluation in scenes with diverse complexity. J Electri Comp Eng 2012:1–16
Borrini M (2007) Archeologia forense. Metodo e tecniche per il recupero dei resti umani: compendio per l'investigazione scientifica. Lo Scarabeo, Bologna
Boyle S, Galloway A, Mason R (1997) Human aquatic taphonomy in the Monterey Bay area. In: Forensic taphonomy: the postmortem fate of human remains. Available at: https://scholar.google.com/scholar?start=60&q=forensic+taphonomy&hl=fr&as_sdt=0,5#1 Accessed 20 May 2016
Brain CK (1981) The hunters or the hunted? An introduction to African cave taphonomy. University of Chicago Press, Chicago
Breuckmann B, Cabal PA, Mélard N, Ontanón R, Pastoors A, Mayoloni LCT, Vega PÁF, Weniger GC (2009) Surface scanning-new perspectives for archaeological data management and methodology. Comp Appl Archaeol:42–47
Bristow J, Simms Z, Randolph-Quinney PS (2011) Taphonomy. In: B S, Ferguson E (eds) Forensic anthropology 2000–2010. CRC Press, Boca Raton, pp 279–318
Brody RH, Edwards HGM, Pollard AM (2001) A study of amber and copal samples using FT-Raman spectroscopy. Spectrochim Acta A 57:1325–1338
Brooks ST, Brooks RH (1984) Problems of burial exhumation, historical and forensic aspects. In: Rathbun TA, Buikstra JE (eds) Human identification. Case studies in forensic anthropology. Charles C. Thomas, Springfield, pp 64–86
Brooks C, MacDonald C (2014) Helicopter crashes into water: warning time, final position, and other factors affecting survival. Aviat Space Environ Med 85(4):440–444. Available at: http://www.ingentaconnect.com/content/asma/asem/2014/00000085/00000004/art00008 Accessed 20 May 2016
Bruinsma GJN, Johnson SD (2018) The Oxford handbook of environmental criminology. Oxford University Press, Oxford
Buchan M, Anderson G (2001) Time since death: a review of the current status of methods used in the later postmortem interval. J Can Soc Forensic Sci 34(1):1–22. Available at: http://www.tandfonline.com/doi/abs/10.1080/00085030.2001.10757514 Accessed 20 May 2016
Buck SC (2003) Searching for graves using geophysical technology: field tests with ground penetrating radar, magnetometry and electrical resistivity. J Forensic Sci 48:5–11
Buikstra JE, Ubelaker DH (1994) Standards for data collection from human skeletal remains. Fayetteville: Arkansas Archeological Survey Research Series No. 44
Byard RW (2014) Unusual patterned skin lesions caused by postmortem fungal activity. Forensic Sci Med Pathol 10:651–653. https://doi.org/10.1007/s12024-014-9589-9
Byard R, James R, Gilbert J (2002) Diagnostic problems associated with cadaveric trauma from animal activity. Am J Forensic Med Pathol 23(3):238–244. Available at: http://journals.lww.com/amjforensicmedicine/Abstract/2002/09000/Diagnostic_Problems_Associated_with_Cadaveric.6.aspx Accessed 21 May 2016
Byard R, James R, Heath K (2006) Recovery of human remains after shark attack. Am J Forensic Med Pathol 27(3):256–259. Available at: http://journals.lww.com/amjforensicmedicine/Abstract/2006/09000/Recovery_of_Human_Remains_After_Shark_Attack.4.aspx Accessed 20 May 2016

Campobasso CP et al (2006) Pink teeth in a series of bodies recovered from a single shipwreck. Am J Forensic Med Pathol 27(4):313–316. Available at: http://content.wkhealth.com/linkback/openurl?sid=WKPTLP:landingpage&an=00000433-200612000-00007 Accessed 16 Aug 2016
Capuano E (2010) Milioni di lingotti di tungsteno dorati al posto dei lingotti d'oro vero. http://old.nexusedizioni.it/index2.php?option=com_content&do_pdf=1&id=1882
Carandini A (1991) Storie dalla Terra. Manuale di scavo archeologico. Einaudi, Torino
Carniel S et al (2002) Tracking the drift of a human body in the coastal ocean using numerical prediction models of the oceanic, atmospheric and wave conditions. Sci Justice 42(3):143–151. Available at: http://www.sciencedirect.com/science/article/pii/S1355030602718194 Accessed 21 May 2016
Casamatta D, Verb R (2000) Algal colonization of submerged carcasses in a mid-order woodland stream. J Forensic Sci 45:1280–1285. Available at: http://www.astm.org/DIGITAL_LIBRARY/JOURNALS/FORENSIC/PAGES/JFS14880J.htm Accessed 12 Dec 2016
Cassidy NJ, Eddies R, Styles P, Dods S (2011) Combining ground-penetrating radar and ultrasonic survey techniques: new tools for old problems? First Break 29:69–75
Castex D, Blaizot F (2017) Reconstructing the original arrangement, organisation and architecture of burials in archaeology. In: Schotsmans EMJ, Nicholas M, Forbes SL (eds) Taphonomy of human remains: forensic analysis of the dead and the depositional environment. Wiley, Hoboken, pp 277–295
Cattaneo C (2007) Forensic anthropology: developments of a classical discipline in the new millennium. Forensic Sci Int 165:185–193
Cattaneo C, Porta D, Gibelli D, Gamba C (2009) Histological determination of the human origin of bone fragments. J Forensic Sci 54:531–533
Capuano E., (2010) Milioni di lingotti di tungsteno dorati al posto dei lingotti d'oro vero. http://old.nexusedizioni.it/index2.php?option=com_content&do_pdf=1&id=1882
Cavenago S, Moneta B (1980) Gemmologia. Hoepli, Milano
Challis K (2004) Trent valley geoarchaeology 2002 component 2b: LiDAR terrain modelling. York Archaeological Trust (unpublished report)
Challis K (2005) Nether Kellet to pannal gas pipeline: airborne Laser altimetry (LiDAR) analysis, HP vista centre. Birmingham Archaeology (unpublished report)
Challis K (2006) Airborne laser altimetry in alluviated landscapes. Archaeol Prospect 13(2):103–127
Chapman R, Wylie A (2015) Material evidence: learning from archaeological practice. Routledge, New York
Chapman R, Wylie A (2016) Evidential reasoning in archaeology. Bloomsbury Academics, London
Claes J et al (2014) Iso-luminance counterillumination drove bioluminescent shark radiation. Sci Rep 10(4):4328. Available at: http://www.nature.com/srep/2014/140310/srep04328/full/srep04328.html?WT.ec_id=SREP-631-20140311 Accessed 20 May 2016
Clarke D (1973) Archaeology: the loss of innocence. Antiquity 47:6–18
Clarke MA, Worrell MB, Pless JE (1997) Postmortem changes in soft tissues. In: Haglund WD, Sorgs MH (eds) Forensic taphonomy: the postmortem fate of human remains. CRC Press, New York, pp 151–164
Cleland CE (2002) Methodological and epistemic differences between historical science and experimental science. Philos Sci 69:474–496
Cleland CE (2011) Prediction and explanation in historical natural sciences. Br J Philos Sci 62:551–582
Cockle D, Bell L (2015) Human decomposition and the reliability of a "Universal" model for post mortem interval estimations. Forensic Sci Int Available at: http://www.sciencedirect.com/science/article/pii/S037907381500211X Accessed 20 May 2016
Commission d'Enquête (2013) Rapport Final sur l'Accident Survenu le 29 Juin 2009 En mer au large de Moroni (Comores) De l' Airbus A310–324 Immatriculé 7O-ADJ Exploité par la compagnie Yemenia Airways, Available at: https://www.bea.aero/docspa/2009/7o-j090629/pdf/7o-j090629.pdf

Compagno LJ, Dando M, Fowler S (2005) A field guide to the sharks of the world. Harper-Collins, London
Connor MA (2007) Forensic methods: excavation for the archaeologist and investigator. AltaMira Press, Lanham
Connor M, Baigent C, Hansen ES (2017) Testing the use of pigs as human proxies in decomposition studies. J Forensic Sci. https://doi.org/10.1111/1556-4029.13727
Cook R, Evett IW, Jackson G, Jones PJ, Lambert JA (1998a) A model for case assessment and interpretation. Sci Justice 38:151–156
Cook R, Evett IW, Jackson G, Jones PJ, Lambert JA (1998b) A hierarchy of propositions: deciding which level to address in casework. Sci Justice 38:231–239
Cook R, Evett IW, Jackson G, Jones PJ, Lambert JA (1999) Case pre-assessment and review in a two-way transfer case. Sci Justice 39:103–111
Coulombeix A, Georges P, Nolot F, Portat E, Richebé J (2014) Archéologie forensique. Gendarmes et archéologues sur une scène de crime. Archéologia 523:68–72
Cox M, Flavel AA, Hanson I, Laver J, Wessling R (2008) The scientific investigation of mass graves. Towards protocols and standard operating procedures. Cambridge University Press, Cambridge
Crist TA (2001) Bad to the bone?: historical archaeologists in the practice of forensic science. Hist Archaeol 35(1):1–6
Crossland Z (2013) Evidential regimes in forensic archaeology. Annu Rev Anthropol 42:121–137
Crowder CM, Andronowski JM, Dominguez VM (2018) Bone histology as an integrated tool in the process of human identification. In: Latham K, Bartelink E, Finnegan M (eds) New perspectives in forensic human skeletal identification. Elsevier, London, pp 201–213
Cubero C (2013) Recomendaciones para la práctica medico forense en supuestos de sustracción de recién nacidos. In: González Fernández J, Gutiérrez-Hoyos A (eds) Décimo Curso de Patología Forense. Gobierno de La Rioja, Logroño, pp 127–139
Cummaudo M, Guerzoni M, Gibelli D, Cigada A, Obertovà Z, Ratnayake M, Poppa P, Gabriel P, Ritz-Timme S, Cattaneo C (2014) Towards a method for determining age ranges from faces of juveniles on photographs. Forensic Sci Int 239:107.e1–107.e7
Cunha E, Cattaneo C (2006) Forensic anthropology and forensic pathology: the state of the art. In: Schmitt A, Cunha E, Pinheiro J (eds) Forensic anthropology and medicine: complementary science from recovery to cause of death. Humana Press, Totowa, pp 39–53
Dabbs GR, Bytheway JA, Connor M (2017) Comparing the scoring of human decomposition from digital images to scoring using on-site observations. J Forensic Sci 62(5):1292–1296. Available at:. https://doi.org/10.1111/1556-4029.13409 Accessed 9 Feb 2017
Dautartas AM, Jantz LM, Vidoli GM, Steadman DW (2015) A multidisciplinary validation study of non-human animal models for forensic decomposition research: a time series approach. In: Proceedings of the 67th annual scientific meeting of the American academy for forensic sciences, Orlando
Davis O (2012) Processing and working with LiDAR data in ArcGIS: a practical guide for archaeologists. The Royal Commission on the Ancient and Historical Monuments of Wales, Aberystwyth
Dawid AP, Evett IW (1997) Using a graphical method to assist the evaluation of complicated patterns of evidence. J Forensic Sci 42:226–231
De Leeuwe R, Groen WJM (2017) A Taphonomic study based on observations of 196 exhumations and 23 clandestine burials. In: Schotsmans EMJ, Nicholas M, Forbes SL (eds) Taphonomy of human remains: forensic analysis of the dead and the depositional environment. Wiley, Hoboken, pp 394–401
De Souza T (2009) Ground penetrating radar for forensics. CINDE J 30(4):16–17
Di Maggio RM (2017) Geological materials on the crime scene. In: Di Maggio RM, Barone PM (eds) Geoscientists at crime scene. A companion to forensic geoscience. Springer, Cham, pp 11–30. ISBN:978- 3-319-58048-7

Di Maggio RM, Barone PM (2017) Geoscientists at crime scenes. A companion to forensic geoscience, soil forensics series. Springer International Publishing, Cham. https://doi.org/10.1007/978-3-319-58048-7 ISBN: 978-3-319-58047-0

Di Maggio RM, Magni P (2016) Geologia forense. La falsificazione dei materiali geologici preziosi. Il Penalista (ISSN 2464–9635), Portali Tematici Giuffrè, A. Giuffrè Editore, Milano

Di Maggio RM, Barone PM, Pettinelli E, Mattei E, Lauro SE, Banchelli A (2013) Geologia forense. Introduzione alle geoscienze applicate alle indagini giudiziarie. Dario Flaccovio Editore, Palermo, p 320

Di Maggio RM, Donnelly LJ, Al Nuaimi KS, Barone PM et al (2017) Global developments in forensic geology. Episodes 2(40):120–131. https://doi.org/10.18814/epiiugs/2017/v40i2/017014

Dickson G, Poulter R, Maas E (2011) Marine bacterial succession as a potential indicator of postmortem submersion interval. Forensic Sci Int 209(1–3):1–10. Available at: http://www.sciencedirect.com/science/article/pii/S0379073810004755 Accessed 20 May 2016

Dirkmaat DC (ed) (2012) A companion to forensic anthropology. Wiley-Blackwell, Chichester

Dirkmaat DC, Cabo LL (2012) Forensic anthropology: embracing the new paradigm. In: Dirkmaat DC (ed) A companion to forensic anthropology. Wiley-Blackwell, Chichester, pp 3–40

Dirkmaat DC, Cabo LL (2016) Forensic archaeology and forensic taphonomy: basic considerations on how to properly process and interpret the outdoor forensic scene. Acad Forensic Pathol 6:439–454

Dirkmaat DC, Cabo LL, Ousley SD, Symes SA (2008) New perspectives in forensic anthropology. Yearb Phys Anthropol 51:33–52

Dirks PHGM, Berger LR, Roberts EM, Kramers JD, Hawks J, Randolph-Quinney PS, Elliott M, Musiba CM, Churchill SE, de Ruiter DJ et al (2015) Geological and taphonomic context for the new hominin species *Homo naledi* from the Dinaledi Chamber, South Africa. eLife 4:e09561

Ditrich H (2015) Cognitive fallacies and criminal investigations. Sci Justice 55:155–159

Dittmar JM (2017) "Cut to the bone": the enhancement and analysis of skeletal trauma using scanning electron microscopy. In: Errickson D, Thompson T (eds) Human remains: another dimension. Academic Press, London, pp 45–56

Domínguez-Rodrigo M, Diez-Martín F, Yravedra J, Barba R, Mabulla A, Baquedano E, Uribelarrea D, Sánchez P, Eren MI (2014) Study of the SHK Main Site faunal assemblage, Olduvai Gorge, Tanzania: implications for Bed II taphonomy, paleoecology, and hominin utilization of megafauna. Quat Int 322–323:153–166

Domínguez-Rodrigo M, Cobo-Sánchez L, Yravedra J, Uribelarrea D, Arriaza C, Organista E, Baquedano E (2017) Fluvial spatial taphonomy: a new method for the study of post-depositional processes. Archaeol Anthropol Sci:1–21

Doneus M (2013) Openness as visualisation technique for interpretive mapping of airborne Lidar derived digital terrain models. Remote Sens 5:6427–6442

Doneus M, Verhoeven G, Fera M, Briese C, Kucera M, Neubauer W (2011) From deposit to point cloud–a study of low-cost computer vision approaches for the straightforward documentation of archaeological excavations. Geoinformatics FCE CTU 6:81–88

Donnelly LJ (2003) The applications of forensic geology to help the police solve crimes. Eur Geol J Eur Fed Geol 16:8–12

Donnelly LD (2008) Subsidence and associated ground movements on the Pennines, northern England. Q J Eng Geol Hydrol 41:315–332

Donnelly LJ (2013a) The applications of forensic geology and geophysics for police and law enforcement ground searches. Second international conference on engineering geophysics. United Arab Emirates University, Al Ain City, pp 82–86, Keynote Presentation

Donnelly LJ (2013b) The design and implementation of a high-assurance forensic geology and police search following the discovery of the Staffordshire (Anglo Saxon) Gold Hoard. In: Pirrie D, Dawson L, Ruffell A (eds) Environmental and criminal Geoforensics, vol 384. Geological Society Special Publication, London, pp 195–208

Donnelly LJ, Harrison M (2010) Development of geoforensic strategy & methodology to search the ground for an unmarked burial or concealed object. Emergency Global Barclay Media Limited, Custom Print, Manchester, pp 30–35

Donnelly L, Harrison M (2013) Geomorphological and geoforensic interpretation of maps, aerial imagery, conditions of diggability and the colour-coded RAG prioritization system in searches for criminal burials. In: Pirrie D, Ruffell A, Dawson LA (eds) Environmental and criminal geoforensics. Geological society special publication no. 384, pp 173–194

Donnelly LJ, Harrison M (2017) Ground searches for graves and buried targets related to homicide, terrorism and organised crime. In: Donnelly LJ. (Guest Editor) 2017. Forensic geology themed issued. Episodes. J Int Union Geol Sci:106–117

Dror IE et al (2017) Letter to the editor-the bias snowball and the bias cascade effects: two distinct biases that may impact forensic decision making. J Forensic Sci Available at: http://doi.wiley.com/10.1111/1556-4029.13496 Accessed 14 April 2017

Duband S, Forest F, Clemenson A (2011) Postmortem injuries inflicted by crawfish: morphological and histological aspects. Forensic Sci Int 206(1):e49–e51. Available at: http://www.sciencedirect.com/science/article/pii/S0379073810003944 Accessed 20 May 2016

Ducrettet F, Georges P, Nolot F, Richebé J, Schuliar Y (2013) Organisation, méthodes et recherches en archéologie criminalistique à l'Institut de recherche criminelle de la gendarmerie 006Eationale. A propos de 20 ans d'expérience. Rev Méd Légale 4:27–37

Duday H (2009) The archaeology of the dead: lectures in archaeothanatology. Oxbow Books, Oxford

Duday H, Guillon M (2006) Understanding the circumstances of decomposition when the body is skeletonized. in forensic anthropology and medicine. In: Schmitt A, Cunha E, Pinheiro J (eds) Complementary sciences from recovery to cause of death. Humana Press, Totowa, pp 117–157

Duday H, Courtaud P, Crubézy E, Sellier P, Tillier AM (1990) L'Anthropologie de terrain: reconnaissance et interprétation des gestes funéraires. Bull Mém Soc Anthropol Paris 2:29–50

Duday H, Le Mort F, Tillier A (2014) Archaeothanatology and funeral archaeology. Application to the study of primary single burials. Anthropologie 52(3):235–346

Duhig C, Martinsen N (2007) Many layers of taphonomy: dismemberment and other body processing. In: Brickley M, Ferllini R (eds) Forensic anthropology: case studies from Europe. Charles C Thomas Ltd., Springfield, pp 86–98

Duijst W, Passier C (2016) Lost and found: the identification process of human bodies recovered from the North Sea. J Forensic Legal Med 41:15–20. Available at: http://www.sciencedirect.com/science/article/pii/S1752928X16300087 Accessed 20 May 2016

Dumser T, Türkay M (2008) Postmortem changes of human bodies on the Bathyal Sea floor—two cases of aircraft accidents above the open sea. J Forensic Sci 53(5):1049–1052. Available at: http://onlinelibrary.wiley.com/doi/10.1111/j.1556-4029.2008.00816.x/full Accessed 20 May 2016

Dupras TL, Schultz JJ, Wheeler SM, Williams LJ (2011) Forensic recovery of human remains: archaeological approaches. CRC Press, Boca Raton

Ebbesmeyer C, Haglund W (1994) Drift trajectories of a floating human body simulated in a hydraulic model of Puget Sound. J Forensic Sci 39(1). Available at: http://www.astm.org/DIGITAL_LIBRARY/JOURNALS/FORENSIC/PAGES/JFS13590J.htm). Accessed 16 Nov 2016

Ebbesmeyer C, Haglund W (2002) Floating remains on Pacific Northwest waters. In: Haglund W, Sorg M (eds) Advances in forensic Taphonomy: method, theory, and archaeological perspectives. CRC Press, Boca Raton, pp 219–240. Available at: https://books.google.com/books?hl=fr&lr=&id=RHE6DAAAQBAJ&oi=fnd&pg=PA219&dq=ebbesmeyer+haglund+2002+floating+remains+in+pacififc&ots=FuiTKo023H&sig=YmU8vP3Qb3jMu2A1onihJkgUwqc Accessed 16 Nov 2016

Efremov IA (1940a) Tafonomiya – novaya otrasl paleontologii // Izvestiya AS USSR. Seriya biologicheskaya. № 3. S. 405–413

Efremov IA (1940b) Taphonomy: new branch of paleontology // pan-American geologist Vol. 74. P. 81–93

Efremov IA (1950) Tafonomiya i geologicheskaya letopis. Trudy paleontologicheskogo institute. Tom XXIV. M.: Izvo AS USSR. In: 179 s

Ellingham STD, Perich P, Tidball-Binz M (2017) The fate of human remains in a maritime context and feasibility for forensic humanitarian action to assist in their recovery and identification. Forensic Sci Int Available at: http://linkinghub.elsevier.com/retrieve/pii/S0379073817302980 Accessed 12 Sept 2017

Ellwood BB, Owsley DW, Ellwood SH, Mercado-Allinger PA (1994) Search for the grave of the hanged Texas gunfighter, William Preston Longley. Hist Archaeol 28:94–112

ENFSI (2015) Guideline for evaluative reporting in forensic science: Strengthening the Evaluation of Forensic Results across Europe (STEOFRAE)

English Heritage (2011) 3D laser scanning for heritage. https://www.english-heritage.org.uk/publications/3d-laser-scanning-heritage2/

Engovatova AV, Mednikova MB, Radyush OA, Pilipenko AS, Shvedchikova TYu, Reshetova IK, Vasilyeva YeYe (2014) Kompleksnye mezhdistsiplinarnye issledovaniya na objekte kulturnogo naslediya federalnogo znacheniya 'Mogila Ermolova Alekseya Petrovicha (1772–1861)' v g. Orle // KSIA. 2014. Vyp. 236. S.313–320

Éric C, Georges P, et Buton Chr (forthcoming) Recherche de corps enfoui illégalement en milieu boisé : mise en place d'un protocole en France, RIPTS

Errickson D (2017) Shedding light on skeletal remains: the use of structured light scanning for 3D archiving. In: Errickson D, Thompson T (eds) Human remains: another dimension. Academic Press, London, pp 93–101

Errickson D, Thompson TJ, Rankin BW (2014) The application of 3D visualization of osteological trauma for the courtroom: a critical review. J Forensic Radiol Imaging 2(3):132–137

Eulitz M, Reiss G (2015) 3D reconstruction of SEM images by use of optical photogrammetry software. J Struct Biol 191(2):190–196

Evett IW (2015) The logical foundations of forensic science: towards reliable knowledge. Philos Trans B 370:20140263

Evett IW, Jackson G, Lambert JA (2000) More on the hierarchy of propositions: exploring the distinction between explanations and propositions. Sci Justice 40:3–10

Evin A, Souter T, Hulme-Beaman A, Ameen C, Allen R, Viacava P (2016) The use of close-range photogrammetry in zooarchaeology: creating accurate 3D models of wolf crania to study dog domestication. J Archaeol Sci Rep 9:87–93

Evis L (2016) Forensic archaeology: the application of comparative excavation methods and recording systems. Archaeopress Archaeology, Oxford

Falkingham PL (2012) Acquisition of high resolution three-dimensional models using free open-source photogrammetric software. Palaeontol Electron 15(1) 1T: 15

Farrugia A, Ludes B (2011) Diagnostic of drowning in forensic medicine. In: Vieira DN (ed) Forensic medicine-from old problems to new challenges. InTech. Available at: http://cdn.intechopen.com/pdfswm/19161.pdf, Rijeka Accessed 20 May 2016

Fenning PJ, Donnelly LJ (2004) Geophysical techniques for forensic investigation. In: Pye K, Croft D (eds) Forensic geoscience: principles, techniques and applications. Special publication of the geological society of London, vol 232, pp 11–20

Ferllini R (2002) Silent witness. Grange Books, Rochester

Fisher BAJ, Fisher DR (2012) Techniques of crime scene investigation, 8th edn. CRC Press, Boca Raton

Fondebrider L (2016) The application of forensic anthropology to the investigation of cases of political violence: perspectives from South America. In: Blau S, Ubelaker DH (eds) Handbook of forensic anthropology and archaeology. Left Coast Press, Walnut Creek, pp 67–75

Forbes S, Stuart B (2004) A preliminary investigation of the stages of adipocere formation. J Forensic Sci 49(3). Available at: https://www.researchgate.net/profile/Ian_Dadour/publication/8535895_A_preliminary_investigation_of_the_stages_of_adipocere_formation/links/02e7e5192dc9a0fc5c000000.pdf). Accessed 20 May 2016

Forbes S, Stuart B, Dent B (2002) The identification of adipocere in grave soils. Forensic Sci Int 127(3):225–230. Available at: http://www.sciencedirect.com/science/article/pii/S0379073802001275 Accessed 20 May 2016

Forbes S, Stuart B, Dent B (2005) The effect of the method of burial on adipocere formation. Forensic Sci Int 154(1):44–52. Available at: http://www.sciencedirect.com/science/article/pii/S0379073804006358 Accessed 20 May 2016

Forbes S, Wilson M, Stuart B (2011) Examination of adipocere formation in a cold water environment. Int J Legal Med 125(5):643–650. Available at: http://link.springer.com/article/10.1007/s00414-010-0460-6 Accessed 20 May 2016

Frailing K, Wood Harper D (2015) Fundamentals of criminology. New dimensions, 2nd edn. Carolina Academic Press, Durham

France DL, Griffin TJ, Swanburg JG, Lindemann JW, Davenport GC, Trammell V, Travis CT, Kondratieff B, Nelson A, Castellano K, Hopkins D, Adair T (1997) Necrosearch revisited: further multidisciplinary approaches to the detection of clandestine graves. In: Haglund WD, Sorg MH (eds) Forensic taphonomy: the postmortem fate of human remains. CRC Press, Boca Raton, pp 497–509

Gadig O, Gomes U (2002) First report on embryos of *Isistius brasiliensis*. J Fish Biol 60(5):1322–1325. Available at: http://onlinelibrary.wiley.com/doi/10.1111/j.1095-8649.2002.tb01723.x/full Accessed 21 May 2016

Gaffney C, Gaffney V, Neubauer W, Baldwin E, Chapman H, Garwood P, Moulden H, Sparrow T, Bates C, Löcker K, Hinterleitner A, Trinks I, Nau E, Zitz T, Floery S, Verhoeven G, Doneus M (2012) The Stonehenge hidden landscapes project. Archaeol Prospect 19:147–155

Gage JD, Tyler PA (1991) In: Gage JD, Tyler PA (eds) Deep-sea biology: a natural history of organisms at the deep-sea floor. Cambridge University Press, Cambridge

Galantucci LM, Pesce M, Lavecchia F (2015) A stereo photogrammetry scanning methodology, for precise and accurate 3D digitization of small parts with sub-millimetre sized features. CIRP Ann Manuf Technol 64(1):507–510

Galeazzi F (2016) Towards the definition of best 3D practices in archaeology: assessing 3D documentation techniques for intra-site data recording. J Cult Herit 17:159–169

Galloway A (1997) The process of decomposition: a model from the Arizona-Sonoran desert. In: Forensic taphonomy: the postmortem fate of human remains. Available at: https://scholar.google.com/scholar?start=60&q=forensic+taphonomy&hl=fr&as_sdt=0,5#2 Accessed 20 May 2016

Gardner RM (2012) Practical crime scene processing and investigation. CRC Press, Boca Raton

Geiling JM, Marín-Arroyo AB (2015) Spatial distribution analysis of the lower Magdalenian human burial in El Mirón Cave (Cantabria, Spain). J Archaeol Sci 60:47–56

Georges P (2009) L'archéologue a-t-il sa place sur la scène de crime ? Localiser, fouiller, prélever les vestiges criminalistiques enfouis. De l'expérience anglo-saxonne au savoir-faire français, Mémoire du DU de Criminalistique, Sous la direction de MM. Phd M..Guillon (Inrap) et du Dr Ph. Charlier (Garches). – Paris 5, Univ. R. Descartes

Georges P (2016) L'archéologie forensique en France : une discipline à l'assaut du temps, dans Du silex au goblet en plastique. Réflexions sur les limites chronologiques de l'archéologie, Bordeaux, 2016, pp 61–90

Georges P, Ducrettet F, Nolot F, Richebé J, Schuliar Y (2012) Problématiques et règles d'emploi criminalistiques des engins de chantier pour la recherche de cadavres enfouis. L'expérience de l'Institut de recherche criminelle de la Gendarmerie nationale (France). Rev Int Criminol Police Tech Sci 3(1):363–374

Georges P, Schuliar Y, Ducrettet F, Nolot F, Richebé J (2013) Les gendarmes et l'archéologie : expertises et recherches de l'IRCGN. Archéothéma 27:6–17

Gerdau- Radonic K (2012) Archaeological insights into the disarticulation pattern of a human body in a sitting/squatting position. Proc Ann Conf Brit Assoc Biol Anthropol Osteoarchaeol 12:151–160

Giacosa G (2017) False and plated coins. Moruzzi Numismatica http://www.moruzzi.it

Giarin M (2012) Pietre preziose. Miriam Giarin E-book

Gibbon G (1989) Explanation in archaeology. Blackwell, Oxford

Gifford DP (1981) Taphonomy and palaeoecology: a critical review of archaeology's sister disciplines. In: Schiffer MB (ed) Advances in archaeological method and theory, vol 4. Academic, New York, pp 365–438

Gifford-Gonzalez D (1989) Modern analogues: developing an interpretive framework. In: Bonnichsen R, Sorg MH (eds) Bone modification. University of Maine Center for the Study of First Americans, Orono, pp 43–52

Gill JR, Rainwater CW, Adams BJ (2009) Santería and Palo Mayombe: skulls, mercury, and artifacts. J Forensic Sci 54:1458–1462

Gill-King H (2009) An incidental finding. In: Steadman DW (ed) Hard evidence: case studies in forensic anthropology, 2nd edn. Prentice Hall, Upper Saddle, pp 47–51

Gould RA (2007) Disaster archaeology. University of Utah Press, Salt Lake City

Graham S, Parkinson C, Chahine M (2010) The water cycle: feature articles. Available at: https://earthobservatory.nasa.gov/Features/Water/page1.php. Accessed 9 Feb 2018

Griffith SJ, Thompson CEL (2017) The use of laser scanning for visualization and quantification of abrasion on water-submerged bone. In: Errickson D, Thompson T (eds) Human remains: another dimension. Academic Press, London, pp 103–122

Groen WJM (2015) Forensic archaeology: the European collaboration. In: Groen WJM, Márquez-Grant N, Janaway RC (eds) Forensic archaeology. A global perspective. Wiley-Blackwell, Oxford, pp 207–211

Groen WJM, Berger CEH (2017) Crime scene investigation, archaeology and taphonomy: reconstructing activities at crime scenes. In: Schotsmans EMJ, Márquez-Grant N, Forbes SL (eds) Taphonomy of human remains: forensic analysis of the dead and the depositional environment. Wiley, Oxford, pp 476–494

Groen WJM, Márquez-Grant N, Janaway R (2015) Forensic archaeology: a global perspective. Wiley-Blackwell, London. ISBN: 978-1-118-74598-4

Gruza E, Mańkowski D (2016) Ewolucja dowodu poszlakowego. In: Wójcikiewicz J, Kwiatkowska-Wójcikiewicz V (eds) Paradygmaty kryminalistyki. Wydawnictwo Instytutu Ekspertyz Sądowych, Kraków

Gruza E, Goc M, Moszczyński J (2008) Kryminalistyka czyli rzecz o metodach śledczych. Wydawnictwa Akademickie i Profesjonalne, Warszawa

Gualdi E, Russo P (2012) La scena del crimine: Ricerca e recupero dei resti umani. Libreria Universitaria Edizioni, Padova

Gunatilake K (1992) Autopsy report case N, pp 92–0650

Guo Q, Pu R, Cheng J (2016) Anomaly detection in hyperspectral remote sensing imagery. Geosciences 6(4)

Gupta N (2013) Local communities, national governments and forensic and archaeological investigations of human right violations. Archaeol J World Archaeol Congress 9:106–131

Hackman L (2016) Forensic anthropology and missing persons investigations. In: Morewitz SJ, C. Study Colls (eds) Handbook of missing persons. Springer International Publishing, Cham, pp 415–425

Haefner J, Wallace J, Merritt R (2004) Pig decomposition in lotic aquatic systems: the potential use of algal growth in establishing a postmortem submersion interval (PMSI). J Forensic Sci Available at: http://www.astm.org/cgi-bin/scholar.cgi?JFS2003283 Accessed 20 May 2016

Haglund W (1993) Disappearance of soft tissue and the disarticulation of human remains from aqueous environments. J Forensic Sci 38(4):806–815. Available at: http://www.contact.astm.org/DIGITAL_LIBRARY/JOURNALS/FORENSIC/PAGES/JFS13476J.htm Accessed 22 May 2016

Haglund WD (2001) Archaeology and forensic death investigations. Hist Archaeol 35(1):26–34

Haglund W, Sorg M (1996) In: Haglund W, Sorg M (eds) Forensic taphonomy: the postmortem fate of human remains. CRC Press, Boca Raton. Available at: https://scholar.google.com/scholar?start=10&q=forensic+taphonomy&hl=fr&as_sdt=0,5#0 Accessed 20 May 2016

Haglund WD, Sorg MD (1997) Forensic taphonomy: the postmortem fate of human remains. CRC Press, Boca Raton

Haglund W, Sorg M (2001) Human remains in water environments. In: Haglund W, Sorg M (eds) Advances in forensic taphonomy: method, theory, and archaeological perspectives. CRC Press. Available at: https://books.google.com/books?hl=fr&lr=&id=5q1rg9Xl9CEC&oi=fnd&pg=PA201&dq=the+taphonomy+of+human+remains+in+glacial+environments&ots=Ss4plYckgD&sig=kxa8qSgKoGlAA5eb4OkXQ_m6Sxk, Boca Raton Accessed 20 May 2016

Häkkänen H, Hurme K, Liukkonen M (2007) Distance patterns and disposal sites in rural area homicides committed in Finland. J Investig Psychol Offender Profiling 4:181–197

Hanausek T (2005) Kryminalistyka. Zarys wykładu, Kraków

Harris OJT, Cipolla C (2017) Archaeological theory in the new millennium. Introducing current perspectives. Routledge, London

Harris EC, Brown MR III, Brown GJ (1993) Practices of archaeological stratigraphy. Academic, London

Harrison M, Donnelly LJ (2008) Buried homicide victims: applied geoforensics in search to locate strategies. J Homicide Major Inv 4:71–86

Hayashi T et al (2015) Postmortem wounds caused by cookie-cutter sharks (Isistius species): an autopsy case of a drowning victim. Forensic Sci Med Pathol 11(1):119–121. Available at: http://link.springer.com/article/10.1007/s12024-014-9597-9 Accessed 20 May 2016

Hays WM (2008) Florida's clandestine graves: an anthropological perspective of the dead. Treatises and Dissertations Paper 4177. Florida State University

Healy C et al (2015) Detecting submerged bodies: controlled research using side-scan sonar to detect submerged proxy cadavers. J Forensic Sci 60(3):743–752. Available at: http://onlinelibrary.wiley.com/doi/10.1111/1556-4029.12671/full Accessed 22 May 2016

Heaton V et al (2010) Predicting the postmortem submersion interval for human remains recovered from U.K. waterways. J Forensic Sci 55(2):302–307

Herrmann NP, Devlin JB (2008) Assessment of commingled human remains using a GIS-based approach. In: Byrd JE, Adams BJ (eds) Recovery, analysis, and identification of commingled human remains. Humana Press, New York, pp 257–270

Hess M, Robson S, Ahmadabadian AH (2014) A contest of sensors in close range 3D imaging: performance evaluation with a new metric test object. Int Arch Photogramm Remote Sens Spat Inf Sci 40(5):277

Hester TR, Shafer JJ, Feder KL (2008) Field methods in archaeology, 7th edn. Left Coast Press, Walnut Creek

Higgs N, Pokines J (2013) Marine environmental alterations to bone. In: Pokines JT, Symes SA (eds) Manual of forensic taphonomy. CRC Press, Boca Raton, pp 143–180. Available at: https://scholar.google.com/scholar?start=440&q=forensic+taphonomy&hl=fr&as_sdt=0,5#8 Accessed 20 May 2016

Higgs ND et al (2011a) Bone-boring worms: characterizing the morphology, rate, and method of bioerosion by *Osedax mucofloris* (Annelida, Siboglinidae). Biol Bull 221(3):307–316. Available at: http://www.journals.uchicago.edu/doi/10.1086/BBLv221n3p307 Accessed 26 July 2017

Higgs ND et al (2011b) Evidence of Osedax worm borings in Pliocene (~3 Ma) whale bone from the Mediterranean. Hist Biol:1–9. Available at: http://www.tandfonline.com/doi/abs/10.1080/08912963.2011.621167 Accessed 26 July 2017

Hill A (1980) Early postmortem damage to the remains of some contemporary East African mammals. In: Behrensmeyer AK, Hill AP (eds) Fossils in the making. Vertebrate taphonomy and paleoecology. University of Chicago Press, Chicago, pp 131–152

Hill A (1984) Hyaenas and hominids: taphonomy and hypothesis testing. In: Foley RA (ed) Hominid evolution and community ecology. Academic Press, London, pp 111–128

Hilts C (2013) The kilkenny workhouse mass burials: an archaeology of the Irish potato famine. Curr Archaeol 278:12–17

Historic England (2017) Understanding the archaeology of landscapes, 2nd edn. Natural England, Swindon

Hobbs NB (1986) Mire morphology and the properties and behaviour of some British and foreign peats. Q J Eng Geol 19:7–80

Hobischak N (1997) Freshwater invertebrate succession and decompositional studies on carrion in British Columbia. Simon Fraser University. Available at: http://www.nlc-bnc.ca/obj/s4/f2/dsk3/ftp04/mq24156.pdf Accessed 20 May 2016

Hobischak N, Anderson G (2002) Time of submergence using aquatic invertebrate succession and decompositional changes. J Forensic Sci 47(1):142–151. Available at: http://www.ncbi.nlm.nih.gov/pubmed/12064642 Accessed 13 June 2016

Hochrein MJ (2002) A autopsy of the grave: recognizing, collecting, and preserving forensic geotaphonomic evidence. In: Haglund WD, Sorg MH (eds) Advances in forensic taphonomy: method, theory, and archaeological perspectives. CRC Press, Boca Raton, pp 45–70

Honebrink R et al (2011) First documented attack on a live human by a cookie cutter shark (Squaliformes, Dalatiidae: Isistius sp.). Pac Sci 65(3):365–374. Available at: http://www.bioone.org/doi/abs/10.2984/65.3.365 Accessed 21 May 2016

Houck MM, Crispino F, McAdam T (2012) The science of crime scenes. Academic, Oxford

Howes LM (2015) The communication of forensic science in the criminal justice system: a review of theory and proposed directions for research. Sci Justice 55(2):145–154. Available at: http://www.ncbi.nlm.nih.gov/pubmed/25754001 Accessed 2 Feb 2018

Howland MD, Liss B, Najjar M, Levy TE (2015) GIS-based mapping of archaeological sites with low-altitude aerial photography and structure from motion: a case study from Southern Jordan. Digital Heritage 2:421–424

Hoyos-Padilla M et al (2013) Observation of an attack by a cookiecutter shark (*Isistius brasiliensis*) on a white shark (*Carcharodon carcharias*). Pac Sci 67(1):129–134 Available at: http://www.scopus.com/inward/record.url?eid=2-s2.0-84872743567&partnerID=tZOtx3y1

Huber D, Motta P (2004) Comparative analysis of methods for determining bite force in the spiny dogfish *Squalus acanthias*. J Exp Zool A Comp Exp Biol. Available at: http://onlinelibrary.wiley.com/doi/10.1002/jez.a.20003/abstract Accessed 21 May 2016

Humphreys M et al (2013) Comparison of protocols for measuring and calculating postmortem submersion intervals for human analogs in fresh water. J Forensic Sci Available at: http://onlinelibrary.wiley.com/doi/10.1111/1556-4029.12033/full Accessed 24 Oct 2016

Hunter J, Cox M (2005) Forensic archaeology: advances in theory and practice. Routledge, London

Hunter J, Martin AL (2002) Locating buried remains. In: Hunter J, Roberts C, Martin A (eds) Studies in crime: an introduction to forensic archaeology. Routledge, London, pp 88–100

Hunter J, Simpson B, Sturdy Colls C (2013) Forensic approaches to buried remains. Wiley-Blackwell, Chichester

Inman K, Rudin N (2002) The origin of evidence. Forensic Sci Int Available at: http://www.sciencedirect.com/science/article/pii/S0379073802000312 Accessed 19 Nov 2016

International Organization for Migration (2017) Mediterranean Migrant Arrivals Top 363,348 in 2016; Deaths at Sea: 5,079. Available at: https://www.iom.int/news/mediterranean-migrant-arrivals-top-363348-2016-deaths-sea-5079. Accessed 23 Nov 2017

Introna F, Di Vella G, Campobasso C (2013) Migrant deaths and the Kater Radez I wreck: from recovery of the relict to marine taphonomic findings and identification of the victims. Int J Legal Med 127(4):871–879. Available at: http://link.springer.com/article/10.1007/s00414-012-0807-2 Accessed 20 May 2016

İşcan MY, Steyn M (2013) The human skeleton in forensic medicine, 3rd edn. Charles C. Thomas, Springfield

Jakubowski O (2015), Criminal threats to the national heritage-an analysis of the events of 2014./in:/Santander Art and Culture Law Review, Nr 1/2015

Jakubowski O (2016), Criminal threats to the national heritage-an analysis of the events of 2015./in:/Santander Art and Culture Law Review, Nr 1/2016

Janaway RC (2008) The decomposition of materials associated with buried cadavers. In: Tibbet M, Carter DO (eds) Soil analysis in forensic Taphonomy: chemical and biological effects of buried human remains. CRC Press, Boca Raton, pp 153–202

Jeffares B (2008) Testing times: regularities in the historical sciences. Stud Hist Phil Biol Biomed Sci 39:469–475

Johnson M (2002) Archaeological theory, an introduction. Blackwell Publishers Ltd, Oxford

Johnson M (2010) Archaeological theory: an introduction. Wiley-Blackwell, Oxford

Johnson et al (2010) Rubyspira, new genus and two new species of bone-eating deep-sea snails with ancient habits. Biol Bull 219(2):166–177. Available at: http://www.ncbi.nlm.nih.gov/pubmed/20972261 Accessed 26 July 2017

Johnston B, Warke PA, Ruffell A, McKinley J (2018) Detecting voids within a historical building façade: a comparative study of three high frequency GPR antennas. J Cult Herit 24:7–10

Jones E et al (1998) The fate of cetacean carcasses in the deep sea: observations on consumption rates and succession of scavenging species in the abyssal north-East Atlantic Ocean. Proc R Soc B Biol Sci 265(1401). Available at: http://rspb.royalsocietypublishing.org/content/265/1401/1119.short). Accessed 24 Nov 2016

Jørkov ML, Lynnerup N (2015) Forensic archaeology in Denmark. In: Mike Groen WJ, Márquez-Grant N, Janaway RC (eds) Forensic archaeology: a global perspective. Wiley, Oxford, pp 55–57

Junkins EN, Carter DO (2017) Relationship between human remains, graves and the depositional environment. In: Shotsmans E, Márquez-Grant N, Forbes S (eds) Taphonomy of human remains: forensic analysis of the dead and the depositional environment. Wiley-Blackwell, Chichester, pp 145–154

Kahana T et al (1999) Marine taphonomy: adipocere formation in a series of bodies recovered from a single shipwreck. J Forensic Sci 44(5):897–901. Available at: http://www.ncbi.nlm.nih.gov/pubmed/10486937 Accessed 13 June 2016

Kalacska ME, Bell LS, Arturo Sanchez-Azofeifa G, Caelli T (2009) The application of remote sensing for detecting mass graves: an experimental animal case study from Costa Rica. J Forensic Sci 54(1):159–166

Kassin SM, Dror IE, Kukucka J (2013) The forensic confirmation bias: problems, perspective and proposed solutions. J Appl Res Mem Cogn 2:42–52

Kashani AG, Olsen MJ, Parrish CE, Wilson N (2015) A review of LIDAR radiometric processing: from ad hoc intensity correction to rigorous radiometric calibration. Sensors 15(11):28099–28128

Keiper J, Casamatta D (2001) Benthic organisms as forensic indicators. J N Am Benthol Soc. Available at: http://www.jstor.org/stable/1468325 Accessed 20 May 2016

Keough N, Myburgh J, Steyn M (2017) Scoring of decomposition: a proposed amendment to the method when using a pig model for human studies. J Forensic Sci 62(4):986–993. Available at: http://onlinelibrary.wiley.com/doi/10.1111/1556-4029.13390/full Accessed 25 Feb 2017

Klepinger L (2006) Fundamentals of forensic anthropology. Wiley, Hoboken

Knupfer G, Godfrey G, Hill J (2018) The independent commission for the location of victims' remains in Ireland: a case study – the abduction, murder and secret burial of Victim B. In: Barone PM, Groen M (eds) Multidisciplinary approaches to forensic archaelology. Soil forensics series. Springer International Publishing, Cham

Knüsel CJ (2014) Crouching in fear: terms of engagement for funerary remains. J Soc Archaeol 14(1):26–58

Knüsel CJ, Robb J (2016) Funerary taphonomy: an overview of goals and methods. J Archaeol Sci Rep 10:655–673

Kola A (2005) Archeologia zbrodni. Oficerowie polscy na cmentarzu ofiar NKWD w Charkowie. UMK, Toruń

Komar DA, Buikstra JE (2008) Forensic anthropology: contemporary theory and practice. Oxford University Press, New York

Komer DA, Buikstra JE (2008) Forensic Taphonomy. Forensic anthropology: contemporary theory and practice. Oxford University Press, New York, pp 189–207
Kranioti E, Paine R (2011) Forensic anthropology in Europe: an assessment of current status and applications. J Anthropol Sci 89:71–92
Kruger A, Randolph-Quinney P, Elliott M (2016) Multimodal spatial mapping and visualisation of Dinaledi chamber and rising star cave. S Afr J Sci 112(5–6):1–11
Lacrêpe C, Georges P, et Buton Chr (forthcoming) Recherche de corps enfoui illégalement en milieu boisé : mise en place d'un protocole en France, RIPTS
Lafaye C (2012) Exemple de contre-insurrection : la fouille opérationnelle en Afghanistan. Rev Historique Armées 268:88–98
Latham K, Bartelink E, Finnegan M (eds) (2018) New perspectives in forensic human skeletal identification. Elsevier, London
Ławecka D (2010) Wstęp do archeologii. Wydawnictwo, Warszawa
Lawler W (1992) Bodies recovered from water: a personal approach and consideration of difficulties. J Clin Pathol. Available at: http://www.ncbi.nlm.nih.gov/pmc/articles/PMC495138/ Accessed 20 May 2016
Ławrynowicz O, Żelazko J (eds) (2015) Archeologia totalitaryzmu. Ślady represji 1939–1956. Instytut Pamięci Narodowej - Komisja Ścigania Zbrodni przeciwko Narodowi Polskiemu, Łódź
Lieskovský T, Ďuračiová R, Karell L (2013) Selected mathematical principles of archaeological predictive models creation and validation in the GIS environment. Interdiscip Archaeol 4:33–46
Litherland S, Marquez-Grant N, Roberts J (2012) Forensic archaeology. In: Marquez-Grant N, Roberts J (eds) Forensic ecology handbook. Wiley-Blackwell, Oxford, pp 24–48
Locard E (1920) L'Enquête Criminelle et les Méthodes Scientifiques. Ernest Flammarion, Paris
Locard E (1930) The analysis of dust and traces. Am J Police Sci 1:279–298, 401–418
Lucas G (2012) Understanding the archaeological record. Cambridge University Press, Cambridge
Lundrigan S, Canter DV (2001a) Spatial patterns of serial murder: an analysis of disposal site location choice. Behav Sci Law 19:595–610
Lundrigan S, Canter DV (2001b) A multivariate analysis of serial murderers´ disposal site location choice. J Environ Psychol 21:423–432
Lyman RL (1994) Vertebrate taphonomy. Cambridge University Press, Cambridge
Lyman RL (2010) What taphonomy is, what it isn't, and why taphonomists should care about the difference. J Taphonomy 8(1):1–16
Lynch T, Cooksey J (2007) Battlefield archaeology. Tempus Publishing, Chalford
Magni P et al (2015) Evaluation of the floating time of a corpse found in a marine environment using the barnacle *Lepas anatifera* L.(Crustacea: Cirripedia: Pedunculata). Forensic Sci Int 247:e6–e10. Available at: http://www.sciencedirect.com/science/article/pii/S0379073814004812 Accessed 20 May 2016
Maile A et al (2017) Toward a universal equation to estimate postmortem interval. Forensic Sci Int 272:150–153. Available at: http://www.sciencedirect.com/science/article/pii/S037907381730018X Accessed 1 Mar 2017
Makino Y, Tachihara K, Ageda S (2004) Peculiar circular and C-shaped injuries on a body from the sea. Am J Forensic Med Pathol 25(2):169–171. Available at: http://journals.lww.com/amjforensicmedicine/Abstract/2004/06000/Peculiar_Circular_and_C_Shaped_Injuries_on_a_Body.17.aspx Accessed 21 May 2016
Marais-Werner A, Myburgh J, Meyer A, Nienaber WC, Steyn M (2017) Decomposition patterns of buried remains at different intervals in the Central Highveld region of South Africa. Med Sci Law 57:115–123, 0025802417705263
Márquez-Grant N (2015) Age estimation in forensic anthropology: perspectives and practical considerations. Ann Hum Biol 42:306–320
Márquez-Grant N, Litherland S, Roberts J (2012) European perspectives and the role of the forensic archaeologist in the UK. In: Dirkmaat DC (ed) A companion to forensic anthropology. Wiley-Blackwell, Chichester, pp 598–625

Márquez-Grant N, García-Rubio A, Harrison K (2016a) El valor de la arqueología y la antropología forense en las operaciones policiales de búsqueda de personas desaparecidas. In: AEAOF (ed) Libro de Ponencia y Comunicaciones. VI Reunión Científica de la Asociación Española de Antropología y Odontología Forense. AEAOF, Madrid, pp 80–90

Márquez-Grant N, Muñoz Villarreal A, López Costas O (2016b) Restos humanos subadultos: cuestiones prácticas en antropología forense. In: Sanabria C (ed) Antropología y Patología Forense de la Muerte: La Investigación Científica y judicial de la Muerte y la Tortura, desde las Fosas Clandestinas, hasta la Audiencia Pública. Forensic Publisher, Bogotá, pp 335–371

Maschner HDG (1996) Middle range theory. In: Fagan BM (ed) The Oxford companion to archaeology. Oxford University Press, New York, p 469

Mateus M, de Pablo H, Vaz N (2013) An investigation on body displacement after two drowning accidents. Forensic Sci Int Available at: http://www.sciencedirect.com/science/article/pii/S0379073813001631 Accessed 20 May 2016

Mateus M, Pinto L, Chambel-Leitão P (2015) Evaluating the predictive skills of ocean circulation models in tracking the drift of a human body: a case study. Aust J Forensic Sci 47(3):322–331. Available at: http://www.tandfonline.com/doi/full/10.1080/00450618.2014.957346 Accessed 3 Feb 2018

Mathys A, Brecko J, Vandenspiegel D, Cammaert L, Semal P (2015) Bringing collections to the digital era three examples of integrated high-resolution digitisation projects. In: Digital heritage, vol 1. IEEE, pp 155–158

Maureille B, Sellier P (1996) Dislocation en ordre paradoxal, momification et décomposition: observations et hypotheses. Bull et Mém de la Société d'anthropologie de Paris, Nouvelle Série 8(3–4):313–327

Mazurkiewicz E, Łysakowski M (2016) Zastosowanie metody georadarowej do badania miejsc pochówków na terenach zurbanizowanych. In: Goc M, Tomaszewski T, Lewandowski R (eds) Kryminalistyka – jedność nauki i praktyki. Przegląd zagadnień z zakresu zwalczania przestępczości. Polskie Towarzystwo Kryminalistyczne, Warszawa

Meadows H (2011) Limb vein analysis for human forensic identification. Unpublished doctoral thesis, University of Dundee

Megyesi MS, Nawrocki SP, Haskell NH (2005) Using accumulated degree-days to estimate the postmortem interval from decomposed human remains. J Forensic Sci 50(3):618–626

Meidl RS, Lovejoy CO (1985) Ectocranial suture closure: a revised method for the determination of skeletal age at death based on the lateral-anterior sutures. Am J Phys Anthropol 68(1):57–66

Meller H (2009) Schlachtfeldarchäologie. Battlefield Archaeology, Halle

Mickleburgh HL, Wescott DJ (2018) Controlled experimental observations on joint disarticulation and bone displacement of a human body in an open pit: implications for funerary archaeology. J Archaeol Sci Rep 20C:158–167

Mitchell S, Booher A (2015) Digitization to realization: the utilization of structure from motion, ESRI ArcScene, and 3D printing to identify taphonomic processes and digitally preserve burial CC-B14. In: Houk BA (ed) The 2015 season of the Chan Chich archaeological project. Texas Tech University, Lubbock, pp 163–174

Mol L, Viles H (2013) Exposing drying patterns: using electrical resistivity tomography to monitor capillary rise in sandstone under varying drying conditions. Environ Earth Sci 68:1647–1659

Molina CM, Pringle JK, Saumett M, Evans GT (2016) Geophysical monitoring of simulated graves with resistivity, magnetic susceptibility, conductivity and GPR in Colombia. S Am Forensic Sci Int 261:106–115

Morgan RM (2017a) Conceptualising forensic science and forensic reconstruction. Part I: a conceptual model. Sci Justice. Available at: http://linkinghub.elsevier.com/retrieve/pii/S1355030617300813 Accessed 26 July 2017

Morgan RM (2017b) Conceptualising forensic science and forensic reconstruction. Part II: the critical interaction between research, policy/law and practice. Sci Justice. Available at: http://linkinghub.elsevier.com/retrieve/pii/S1355030617300825 Accessed 26 July 2017

Morgan OW et al (2006) Mass fatality management following the South Asian tsunami disaster: case studies in Thailand, Indonesia, and Sri Lanka. PLoS Med 3(6). Available at: http://journals.plos.org/plosmedicine/article?id=10.1371/journal.pmed.0030195). Accessed 20 May 2016

Morton RJ, Lord WD (2002) Detection and recovery of abducted and murdered children: behavioral and taphonomic influences. In: Haglund WD, Sorg MF (eds) Advances in forensic taphonomy: method, theory and archaeological perspectives. CRC Press, Boca Raton, pp 151–171

Motta P, Wilga C (2001) Advances in the study of feeding behaviors, mechanisms, and mechanics of sharks. In: Tricas T, Gruber S (eds) The behavior and sensory biology of elasmobranch fishes: an anthology in memory of Donald Richard Nelson. Kluwer Academic PUblishers, Dordrecht, pp 131–156. Available at: http://link.springer.com/chapter/10.1007/978-94-017-3245-1_10 Accessed 21 May 2016

Mulhern DM (2016) Differentiating human from nonhuman skeletal remains. In: Blau S, Ubelaker DH (eds) Handbook of forensic anthropology and archaeology, 2nd edn. Routledge, New York, pp 197–212

Mundorff AZ (2014) Anthropologist-directed triage: three distinct mass fatality events involving fragmentation and commingling of human remains. In: Adams BJ, Byrd JE (eds) Commingled human remains: methods in recovery, analysis, and identification. Academic, Oxford, pp 365–388

Murphy RR et al (2011) Use of remotely operated marine vehicles at Minamisanriku and Rikuzentakata Japan for disaster recovery. In 2011 IEEE International Symposium on Safety, Security, and Rescue Robotics. IEEE, pp. 19–25. Available at: http://ieeexplore.ieee.org/document/6106798/. Accessed 3 Feb 2018

Nakhaeizadeh S et al (2018) Cascading bias of initial exposure to information at the crime scene to the subsequent evaluation of skeletal remains. J Forensic Sci 63(2):403–411

Nelson AJ (1998) Wandering bones: archaeology, forensic science and Moche burial practices. Int J Osteoarchaeol 8(3):192–212

National Oceanic and Atmospheric Administration (2017) How far does light travel in the ocean? US Department of Commerce. Available at: https://oceanservice.noaa.gov/facts/light_travel.html. Accessed 1 March 2018

Nethery K (2004) Non-familial abductions that end in homicide: an analysis of the distance patterns and disposal sites. Simon Fraser University, Burnaby

Newman SE (2015) Applications of reflectance transformation imaging (RTI) to the study of bone surface modifications. J Archaeol Sci 53:536–549

Nobes DC (1999) Geophysical surveys of burial sites: a case study of the Oaro Urupa site. Geophysics 64:357–367

Nobes DC (2000) The search for "Yvonne": a case example of the delineation of a grave using near-surface geophysical methods. J Forensic Sci 45:715–721

Noce V (2008) Kolekcja egoisty. Ośrodek Ochrony Zbiorów Publicznych, Warszawa

Nordby JJ (2002) Is forensic taphonomy scientific? In: Haglund WD, Sorg MH (eds) Advances in forensic taphonomy: method, theory, and archaeological perspectives. CRC Press, Boca Raton, pp 31–45

Notter S, Stuart B (2012) The effect of body coverings on the formation of adipocere in an aqueous environment. J Forensic Sci Available at: http://onlinelibrary.wiley.com/doi/10.1111/j.1556-4029.2011.01943.x/full Accessed 20 May 2016

O'Brien T, Kuehner A (2007) Waxing grave about adipocere: soft tissue change in an aquatic context. J Forensic Sci Available at: http://onlinelibrary.wiley.com/doi/10.1111/j.1556-4029.2006.00362.x/full Accessed 20 May 2016

Obledo MN (2009) Forensic archaeology in criminal and civil cases. Forensic Mag 6:3

Odes EJ, Parkinson AH, Randolph-Quinney PS, Zipfel B, Jakata K, Bonney H, Berger LR (2017) Osteopathology and insect traces in the *Australopithecus africanus* skeleton StW 431. S Afr J Sci 113(1/2):20160143

Olson EC (1980) Taphonomy: its history and role in community ecology. In: Behrensmeyer AK, Hill AP (eds) Fossils in the making. Chicago University Press, Chicago, pp 5–19

Ortiz A, Chambon P, Molist M (2013) "Funerary bundles" in the PPNB at the archaeological site of Tell Halula (middle Euphrates valley, Syria): analysis of the taphonomic dynamics of seated bodies. J Archaeol Sci 40:4150–4161

Owings Webb PA, Suchey JM (1985) Epiphyseal union of the anterior iliac crest and medial clavicle in a modern multiracial sample of American males and females. Am J Phys Anthropol 68:457–466

Pampín J, Rodríguez B (2001) Surprising drifting of bodies along the coast of Portugal and Spain. Legal Med. Available at: http://www.sciencedirect.com/science/article/pii/S1344622301000323 Accessed 21 May 2016

Papastamatiou Y, Wetherbee B (2010) Foraging ecology of cookiecutter sharks (*Isistius brasiliensis*) on pelagic fishes in Hawaii, inferred from prey bite wounds. Environ Biol Fish 88(4):361–368. Available at: http://link.springer.com/article/10.1007/s10641-010-9649-2 Accessed 21 May 2016

Paulsen DJ, Robinson MB (2008) Crime mapping and spatial aspects of crime, 2nd edn. Pearson, Harlow

Paz A, Molero JM, Garzon EM, Martinez JA, Plaza A (2010) A new parallel implementation of the RX algorithm for anomaly detection in hyperspectral images. In: Proceedings of the 10th International Conference on Computational and Mathematical Methods in Science and Engineering. CMMSE 2010

Peressinotto MD (2007) Chronologie de la dislocation articulaire du squelette axial et des ceintures au cours de la décomposition du cadavre. Apports à l'analyse des sépultures. Unpublished Ph.D. dissertation, University of Bordeaux

Pesquero MD, Bell LS, Fernández-Jalvo Y (2017) Skeletal modification by microorganisms and their environments. Hist Biol:1–12. Available at: https://www.tandfonline.com/doi/full/10.1080/08912963.2017.1371713 Accessed 10 Feb 2018

Petrik M, Hobischak N, Anderson GS (2004) Examination of factors surrounding human decomposition in freshwater: a review of body recoveries and coroner cases in British Columbia. J Can Soc Forensic Sci 37(1):9–17. Available at: http://www.tandfonline.com/doi/abs/10.1080/00085030.2004.10757565 Accessed 20 May 2016

Pierce C (1989). A critique of middle-range theory in archaeology. Available at: http://webdataworks.com/files/Critique%20of%20Middle%20Range%20Theory%20in%20Archaeology.pdf. Last Accessed 25 June 2018

Plets G, Gheyle W, Verhoeven G, De Reu J, Bourgeois J, Verhegge J, Stichelbaut B (2012) Three-dimensional recording of archaeological remains in the Altai Mountains. Antiquity 86(333):884–897

Pobiner BL, Braun DR (2005) Applying actualism: considerations for future research. J Taphonomy 3:57–65

Pokines J (2018) Differential diagnosis of the Taphonomic histories of common types of forensic osseous remains. J F I 68(1):87–145. Available at: https://www.researchgate.net/profile/James_Pokines/publication/322869412_Differential_diagnosis_of_the_taphonomic_histories_of_common_types_of_forensic_osseous_remains/links/5a735edaa6fdcc53fe146a0c/Differential-diagnosis-of-the-taphonomic-histories-of-co Accessed 8 April 2018

Pokines J, Higgs N (2015) Macroscopic Taphonomic alterations to human bone recovered from marine environments. J F I. Available at: http://search.proquest.com/openview/3bd4b86a6828786a3206e549cf093b2f/1.pdf?pqorigsite=gscholar&cbl=29772 Accessed 20 May 2016

Pokines JT, Symes SA (eds) (2014) Manual of forensic Taphonomy. CRC Press, Boca Raton

Powell KJ (2006) The detection of buried human skeletal remains in the Australian environment. PhD thesis, University of Adelaide, Australia

Pringle JK, Jervis JR (2010) Electrical resistivity survey to search for a recent clandestine burial of a homicide victim, UK. Forensic Sci Int 202:e1–e7

Pringle JK, Ruffell A, Jervis JR, Donnelly L, McKinley J, Hansen J, Morgan R, Pirrie D, Harrison M (2012) The use of geoscience methods for terrestrial forensic searches. Earth Sci Rev 114(1–2):108–123. https://doi.org/10.1016/j.earscirev.2012.05.006

Pringle JK, Jervis JR, Roberts D, Dick HC, Wisniewski KD, Cassidy NJ, Cassella JP (2016) Geophysical monitoring of simulated clandestine graves using electrical and ground penetrating radar methods: 4–6 years. J Forensic Sci 61:309–321

Purcell, M. et al., (2011) Use of REMUS 6000 AUVs in the search for the Air France Flight 447. In OCEANS'11 MTS/IEEE KONA. IEEE, pp. 1–7. Available at: http://ieeexplore.ieee.org/document/6107253/. Accessed 3 Feb 2018

Pye K, Croft D (2004) Forensic geoscience: principles, techniques and applications. Special Publications 232. Geological Society, London

Quick MI (2017) Predicting optimal forensic remote sensing and near-surface detection methods, poster presented at European Meeting on Forensic Archaeology (EMFA) 2017. Rome

Raab LM, Goodyear AC (1984) Middle-range theory in archaeology: a critical review of origins and applications. Am Antiq 49:255–268

Ramirez-Llodra E et al (2010) Deep, diverse and definitely different: unique attributes of the world's largest ecosystem. Biogeosciences 7(9):2851–2899. Available at: http://www.biogeosciences.net/7/2851/2010/ Accessed 6 March 2018

Randolph-Quinney PS (2015) A new star rising: biology and mortuary behaviour of *Homo naledi*. S Afr J Sci 111(9/10):a0122

Rawlins BG, Kemp SJ, Hodgkinson EH, Riding JB, Vane CH, Poulton C, Freeborough K (2006) Potential and pitfalls in establishing the provenance of earth-related samples in forensic investigations. Forensic Sci Int 51:832–845

Rebman A, David E, Sorg MH (2000) The cadaver dog handbook: forensic training and tactics for the recovery of human remains. CRC Press, Boca Raton

Renfrew C, Bahn P (2016) Archaeology theories, methods and practice, 7th edn. Thames and Hudson Ltd, London

Reynolds JM (2011) An introduction to applied and environmental geophysics, 2nd edn. Wiley, Chichester, 711pp

Ribaux O, Baylon A, Roux C, Delémont O, Lock E et al (2010a) Intelligence-led crime scene processing. Part I: forensic intelligence. Forensic Sci Int 195:10–16

Ribaux O, Baylon A, Lock E, Delémont O, Roux C et al (2010b) Intelligence-led crime scene processing. Part II: intelligence and crime scene examination. Forensic Sci Int 199:63–71

Ribéreau-Gayon, A (2014) Unusual patterns of scavenging on the submerged victims of an airplane crash–Forensic anthropologic contribution to the study of marine taphonomy. University College London (UCL)

Ribéreau-Gayon A, Rando C, Schuliar Y et al (2016) Extensive unusual lesions on a large number of immersed human victims found to be from cookiecutter sharks (Isistius spp.): an examination of the Yemenia plane crash. Int J Legal Med:1–10. https://doi.org/10.1007/s00414-016-1449-6

Ribéreau-Gayon A et al (2017) Extensive unusual lesions on a large number of immersed human victims found to be from cookiecutter sharks (Isistius spp.): an examination of the Yemenia plane crash. Int J Legal Med 131(2)

Ribéreau-Gayon A, Carter DO, Regan S (2018) New evidence of predation on humans by cookiecutter sharks in Kauai, Hawaii. Int J Legal Med:1–7. Available at: http://link.springer.com/10.1007/s00414-018-1786-8 Accessed 14 Feb 2018

Richter T, Stock JT, Maher L, Hebron C (2010) An early epipalaeolithic sitting burial from the Azraq oasis. Jordan Antiq 84:321–334

Ritter E, Levine M (2005) Bite motivation of sharks reflected by the wound structure on humans. Am J Forensic Med Pathol Available at: http://journals.lww.com/amjforensicmedicine/Abstract/2005/06000/Bite_Motivation_of_Sharks_Reflected_by_the_Wound.7.aspx Accessed 20 May 2016

Robertson B, Vignaux GA, Berger CEH (2016) Interpreting evidence: evaluating forensic science in the courtroom, 2nd edn. Wiley, Oxford

Roberts J, Márquez-Grant N (2012) Forensic anthropology. In: Márquez-Grant N, Roberts J (eds) Forensic ecology handbook: from crime scene to court. Wiley-Blackwell, Chichester, pp 49–67
Rodriguez W, Bass W (1985) Decomposition of buried bodies and methods that may aid in their location. J Forensic Sci 30(3):836–852
Rogers TL (2009) Skeletal age estimation. In: Blau S, Ubelaker DH (eds) Handbook of forensic anthropology and archaeology. Left Coast Press, Walnut Creek, pp 208–221
Roksandic M (2002) Position of skeletal remains as a key to understanding mortuary behavior. In: Haglund WD, Sorg MH (eds) Advances in forensic Taphonomy: method, theory, and archaeological perspectives. CRC Press, Boca Raton, pp 99–117
Rolandi V, Cavagna S (1996) Procedure e metodi di indagine tradizionali ed avanzati per l'identificazione delle gemme. Hoepli, Milano
Rosen L (1977) The anthropologist as expert witness. Am Anthropol 79:555–578
Rossmo DK (2000) Geographic profiling. CRC Press, Boca Raton
Rottier S (2016) The seated dead: evidence of funerary complexity from the early late bronze age, 14th–12th centuries BCE in France. J Archaeol Sci Rep 10:810–818
Rowlands A, Sarris A (2007) Detection of exposed and subsurface archaeological remains using multi-sensor remote sensing. J Archaeol Sci 34(5):795–803
Rubezhansky AF (1975) Nekotorye voprosy osobennosti osmotra mesta proishestviya pri obnaruzhenii kostnyh ostankov // Sovrmennye voprosy sudebnoi meditsiny I ekspertnoi praktiki. Vyp. 3. Izhevsk S:279–280
Rubezhansky AF (1978) Opredelenie po kostnym ostankam davnostizahoroneniya trupa. M., Medicina. 120 s
Rudwick MJS (1971) Uniformity and progression: reflections on the structure of geological. University of Oklahoma Press, Norman
Ruffell A (2004) Burial location using cheap and reliable quantitative probe measurements. Divers Forensic Anthropol Forensic Sci Int 151:207–211. https://doi.org/10.1016/j.forsciint.2004.12.036
Ruffell A (2005) Under-water scene investigation using ground penetrating radar (GPR) in the search for a sunken jet ski, Northern Ireland. Sci Justice: J Forensic Sci. Available at: http://europepmc.org/abstract/med/17500424 Accessed 11 Nov 2016
Ruffell A, McAllister S (2014) A RAG system for forensic and archaeological searches of burial grounds. Int J Archaeol 3:1–8
Ruffell A, McKinley J (2008a) Geoforensics. Wiley-Blackwell, Chichester. ISBN:978-0-470-05734-6
Ruffell A, McKinley J (2008b) Geoforensics. Wiley Interscience, London, 352pp
Ruffell A, Pringle JK, Forbes S (2014) Search protocols for hidden forensic objects beneath floors and within walls. Forensic Sci Int 237:137–145
Ruffell A, Pringle JK, Cassella JP, Morgan RM, Ferguson M, Heaton VG, Hope C, McKinley JM (2017) The use of geoscience methods for aquatic forensic searches. Earth Sci Rev 117:323–327
Saferstein R (2015) Criminalistics. An introduction to forensic science, 11th edn. Pearson, Harlow
Salmon WC (1992) Explanation in archaeology: an update. In: Embree L (ed) Metaarchaeology. Kluwer Academic Publishers, Dordrecht, pp 243–253
Sanders T (2005) The torso in the Thames: imagining darkest Africa in the United Kingdom. In: Meneley A, Young DJ (eds) Auto-ethnographies: the anthropology of academic practices. Broadview Press, Toronto, pp 126–142
Schiffer MB (1976) Behavioural archaeology. Academic Press, New York
Schiffer MB (1983) Toward the identification of formation process. Am Antiq 48:675–706
Schiffer MB (1987) Formation processes of the archaeological record. University of New Mexico Press, Albuquerque
Schiffer MB (1988) The structure of archaeological theory. Am Antig 53:461–485
Schofield P, Vannan A (2011) Holwick, upper teesdale. Community Survey, Oxford Archaeology North, County Durham (unpublished report)

Schotsmans E, Márquez-Grant N, Forbes S (2017a) Introduction. In: Shotsmans E, Márquez-Grant N, Forbes S (eds) Taphonomy of human remains: forensic analysis of the dead and the depositional environment. Wiley-Blackwell, Chichester, pp 1–8

Schotsmans E, Márquez-Grant N, Forbes S (2017b) Taphonomy of human remains: forensic analysis of the dead and the depositional environment. Wiley-Blackwell, Chichester

Schotsmans E, Georges P, Coulombeix A, Groen WJ (forthcoming) « Forensic archaeology in Europe », Actes du colloque « (Re)lecture archéologique de la justice en Europe médiévale et moderne », Bordeaux 8–10 février 2017, à paraître

Schuliar Y et al (2014) A new tool for coding and interpreting injuries in fatal airplane crashes: the crash injury pattern assessment tool application to the Air France Flight AF447. J Forensic Sci 59(5):1263–1270. Available at: http://onlinelibrary.wiley.com/doi/10.1111/1556-4029.12529/full Accessed 20 May 2016

Schuliar Y, Georges P, Ducrettet F, Nolot F, Richebé J (2015) Forensic archaeology in the French context: the role of the forensic sciences Institute of the French National Gendarmerie. In: Groen DM, Màrquez-Grant N, Janaway R (eds) Forensic archaeology: current trends and future perspectives. Wiley, New York, pp 59–66

Schultz JJ (2012) The application of ground-penetrating radar for forensic grave detection. In: Dirkmaat D (ed) A companion to forensic anthropology, chapter 4. Wiley, London, pp 85–100. https://doi.org/10.1002/9781118255377.ch4

Scott DD, Connor M (2001) The role and future of archaeology in forensic science. Hist Archaeol 35(1):1–6

Scott DD, Connor M, Michael RL (2001) Archaeologists as forensic investigators: defining the role. Society for Historical Archaeology, Tucson

Shark Research Institute (2017) Global Shark Attack File. Available at: http://www.sharkattack-data.com/gsaf/place/united_states_of_america/hawaii. Accessed 4 Oct 2017

Shean B, Messinger L, Papworth M (1993) Observations of differential decomposition on sun exposed v. shaded pig carrion in coastal Washington State. J Forensic Sci. Available at: http://www.astm.org/DIGITAL_LIBRARY/JOURNALS/FORENSIC/PAGES/JFS13492J.htm Accessed 23 Oct 2016

Sherman D, Nordstrom K (1994) Hazards of wind-blown sand and coastal sand drifts: a review. J Coast Res. Available at: http://www.jstor.org/stable/25735603 Accessed 21 May 2016

Shirai S, Nakaya K (1992) Functional morphology of feeding apparatus of the cookie-cutter shark, *Isistius brasiliensis* (Elasmobranchii, Dalatiinae). Zool Sci 9(4):811–821. Available at: http://cat.inist.fr/?aModele=afficheN&cpsidt=4440329 Accessed 21 May 2016

Shirley NR, Wilson RJ, Meadows Jantz L (2011) Cadaver use at the University of Tennessee's anthropological research facility. Clin Anat 24:372–380

Shott MJ (1998) Status and role of formation theory in contemporary archaeological practice. J Archaeol Res 6:299–330

SHOM (2009) Soutien du SHOM au BEA Recherche de l'A330–200 AF447, Available at: https://www.bea.aero/fileadmin/documents/enquetes/vol.af.447/shom.050609.pdf

Simmons T (2002) Taphonomy of a karstic cave execution site at Hrgar, Bosnia-Herzegovina. In: Sorg MH, Haglund WD (eds) Advances in forensic taphonomy: method, theory and archeological perspectives. CRC Press, Boca Raton, pp 263–276

Simmons T (2017) Post-mortem interval estimation: an overview of techniques. In: Schotsmans EMJ, Nicholas M, Forbes SL (eds) Taphonomy of human remains: forensic analysis of the dead and the depositional environment. Wiley, Hoboken, pp 134–142

Simmons T & Heaton V (2013) Postmortem Interval: Submerged Bodies. Wiley Encyclopedia of Forensic Science. Available at: http://onlinelibrary.wiley.com/doi/10.1002/9780470061589.fsa1080/full. Accessed 20 May 2016

Simmons T, Cross P et al (2010a) The influence of insects on decomposition rate in buried and surface remains. J Forensic Sci 55(4):889–892. Available at: http://onlinelibrary.wiley.com/doi/10.1111/j.1556-4029.2010.01402.x/full Accessed 20 May 2016

Simmons T, Adlam R, Moffatt C (2010b) Debugging decomposition data–comparative taphonomic studies and the influence of insects and carcass size on decomposition rate. J Forensic Sci. Available at: http://onlinelibrary.wiley.com/doi/10.1111/j.1556-4029.2009.01206.x/full Accessed 20 May 2016

Simons LM (2010) Guerre di fossili. http://www.nationalgeographic.it/scienza/2010/04/12/news/fossil_wars-3625/

Sławik K, Wasilewska MA (eds) (2003) Niekonwencjonalne metody rozpoznawczo-wykrywcze. Wydawnictwo Hogben, Szczecin

Sledzik P, Mundorff AZ (2016) Forensic anthropology in disaster response. In: Blau S, Ubelaker DH (eds) Handbook of forensic anthropology and archaeology, 2nd edn. Routledge, New York, pp 477–495

Sledzik PS, Dirkmaat D, Mann RW, Holland TD, Mundorff AZ, Adams BJ, Crowder CM, De Paolo F (2009) Disaster victim identification: forensic anthropology in the aftermath of September 11. In: Steadman DW (ed) Hard evidence: case studies in forensic anthropology, 2nd edn. Prentice Hall, Upper Saddle River, pp 289–302

Slizewski A, Semal P (2009) Experiences with low and high cost 3D surface scanners. Quartärs 56:131–138

Slizewski A, Friess M, Semal P (2010) Surface scanning of anthropological specimens: nominal-actual comparison with low cost laser scanner and high-end fringe light projection surface scanning systems. Quartär 57:179–187

Smirnov Yu A (1997) Labirint: Morfologiya prednamerennogo pogrebeniya. M., Vostochnaya literature RAN. 279 s

Smith ME (2011) Empirical urban theory for archaeologists. J Archaeol Method Theory 18:167–192

Smith CR, Glover AG, Treude T et al (2015) Whale-fall ecosystems: recent insights into ecology, paleoecology, and evolution. Annu Rev Mar Sci 7:571–596

Snook B, Cullen RM, Mokros A, Harbort S (2005) Serial murderers' spatial decisions: factors that influence crime location choice. J Investig Psychol Offender Profiling 2:147–164

Sorg MH, Haglund WD (2002) Advancing forensic taphonomy: purpose, theory, and process. In: Sorg MH, Haglund WD (eds) Advances in forensic Taphonomy: method, theory and archeological perspectives. CRC Press, Boca Raton, pp 4–29

Sorg M et al (1997) Forensic Taphonomy in marine contexts. In: Haglund W, Sorg M (eds) Forensic taphonomy: the postmortem fate of human remains. Boca Raton, USA, pp 567–604 Available at: https://books.google.fr/books?hl=en&lr=&id=oNsbKIeZpcQC&oi=fnd&pg=PR15&dq=Sorg,+M.+H.%3B+Dearborn,+J.+H.%3B+Monahan,+E.+I.%3B+Ryan,+H.+F.%3B+Sweeney,+K.+G.%3B+David,+E.+Forensic+Taphonomy+in+Marine+Contexts.+In+Forensic+Taphonomy:+The+Postmortem+Fate+of+H

Sorg MH, Haglund WD, Wren JA (2012) Current research in forensic taphonomy. In: Dirkmaat DC (ed) A companion to forensic anthropology. Wiley-Blackwell Publishing, Chichester, pp 477–498

Steadman DW, Dautartas A, Jantz LM, Mundorff A, Vidoli GM. (2016) A multidisciplinary validation study of nonhuman animal models for human decomposition research. National Institute of Justice Forensic Science Research and Development Symposium. RTI Press Publication No. CP-0002-1607. RTI Press, Research Triangle Park

Stejskal SM (2013) Death, decomposition, and detector dogs: from science to scene. CRC Press, Taylor & Francis, Boca Raton

Stewart TD (1979) Essentials of forensic anthropology. Charles C. Thomas, Springfield

Stock MK, Winburn AP, Burgess GH (2017) Skeletal indicators of shark feeding on human remains: evidence from Florida forensic anthropology cases. J Forensic Sci 62(6):1647–1654. Available at: http://www.ncbi.nlm.nih.gov/pubmed/28464294 Accessed 4 May 2017

Stoel RD, Berger CEH, Kerkhoff W, Matthijssen EJAT, Dror IE (2014) Minimizing contextual bias in forensic casework. In: Strom KJ, Hickman MJ (eds) Forensic science and the Administration of Justice: critical issues and directions. SAGE Publications, Thousand Oaks, pp 67–86

Strongman KB (1992) Forensic applications of ground penetrating radar. In: Pilon J (ed) Ground penetrating radar. Geological Survey of Canada. Paper 90–4, 203–11

Stuart B, Ueland M (2017a) Decomposition in aquatic environments. In: Schotsmans E, Márquez-Grant N, Forbes SL (eds) Taphonomy of human remains: forensic analysis of the dead and the depositional environment. Wiley, Chichester, pp 235–250

Stuart B, Ueland M (2017b) Degradation of clothing in depositional environments. In: Shotsmans E, Márquez-Grant N, Forbes S (eds) Taphonomy of human remains: forensic analysis of the dead and the depositional environment. Wiley-Blackwell, Chichester, pp 120–133

Stuart B et al (2005) Studies of adipocere using attenuated total reflectance infrared spectroscopy. Forensic Sci Med Pathol 1(3):197–201

Stuart B et al (2016) The formation of adipocere in model aquatic environments. Int J Legal Med 130(1):281–286

Suchey JM, Brooks S (1990) Skeletal age determination based on the os pubis: a comparison of the Acsádi-Nemeskéri and Suchey-Brooks methods. Hum Evol 5(3):227–238

Sullivan AP (1978) Inference and evidence in archaeology: a discussion of the conceptual problems. Adv Archeol Method Theory 1:183–222

Swartz BK Jr (1967) A logical sequence of archaeological objectives. Am Antiq 32:487–497

Szumski A (2010) Wykorzystanie taktyki i techniki kryminalistycznej w ramach czynności operacyjno-rozpoznawczych. Oficyna Wydawnicza Arboretum, Wrocław

Tarr H (1954) Microbiological deterioration of fish post mortem, its detection and control. Bacteriol Rev. Available at: https://www.ncbi.nlm.nih.gov/pmc/articles/PMC180782/ Accessed 23 Jan 2017

Taylor D, Ashby K, Winkel K (2002) An analysis of marine animal injuries presenting to emergency departments in Victoria, Australia. Wilderness Environ Med. Available at: http://www.sciencedirect.com/science/article/pii/S1080603202706054. Accessed 23 Oct 2016

Tersigni-Tarrant MTA, Langley NR (2017) Forensic anthropology in the United States: past and present. In: Langley NR, Tersigni-Tarrant MTA (eds) Forensic anthropology: a comprehensive introduction. CRC Press, Boca Raton, pp 3–22

Thali MJ, Braun M, Wirth J, Vock P, Dirnhofer R (2003) 3D surface and body documentation in forensic medicine: 3-D/CAD photogrammetry merged with 3D radiological scanning. J Forensic Sci 48(6):1356–1365

The United Nations High Commissioner for Refugees (2016) Mediterranean Sea: 100 people reported dead yesterday, bringing year total to 5,000. Available at: http://www.unhcr.org/news/briefing/2016/12/585ce804105/mediterranean-sea-100-people-reported-deadyesterday-bringing-year-total.html. Accessed 23 Nov 2017

Thomas TB, Finley SJ, Wilkinson JE, Wescott DJ, Gorski A, Javan GT (2017) Postmortem microbial communities in burial soil layers of skeletonized humans. J Forensic Legal Med 49:43–49

Thompson T, Black S (2006) Forensic human identification: an introduction. CRC Press, Boca Raton

Tibbett M, Carter DO (eds) (2008) Soil analysis in forensic taphonomy: chemical and biological effects of buried human remains. CRC Press, Boca Raton

Tibbett M, Carter DO (2009) Research in forensic taphonomy: a soil based perspective. In: Ritz K, Dawson L, Miller D (eds) Criminal and environmental soil forensics. Springer, Dordrecht

Todd TW (1920) Age changes in the pubic bone: the male Pubic bone. Johann Wolfgang Goethe-Universität, keine Angabe Fachbereich, keine Angabe Institut, Frankfurt am Main, pp 313–314

Tresguerres M, Katz S, Rouse GW (2013) How to get into bones: proton pump and carbonic anhydrase in Osedax boneworms. Proc R Soc Lond B Biol Sci. Available at: http://rspb.royalsocietypublishing.org/content/280/1761/20130625.short Accessed 26 July 2017

Trotter M (1970) Estimation of stature from intact limb long bones. In: Stewart TD (ed) personal identification in mass disasters. Smithsonian Institution, Washington, pp 71–83

Troutman L, Moffatt C, Simmons T (2014) A preliminary examination of differential decomposition patterns in mass graves. J Forensic Sci 59(3):621–626

Trzciński M (2010) Przestępczość przeciwko zabytkom archeologicznym. Problematyka prawno-kryminalistyczna. Wolters Kluwer, Warszawa

Trzciński M (ed) (2013) Archeologia sądowa w teorii i praktyce. Wolters Kluver, Warszawa

Trzciński M (2017) Praktyczne problemy realizacji badań archeologicznych w związku z prowadzeniem poszukiwań ofiar zbrodni komunistycznych, "Kurier Konserwatorski", No. 14

Trzciński M, Jakubowski O (2016) Przestępczość przeciwko dziedzictwu kulturowemu. Diagnoza, zapobieganie, zwalczanie. Katedra Kryminalistyki Wydział Prawa, Administracji i Ekonomii UWr, Wrocław

Tschauner H (1996) Middle-range theory, behavioral archaeology, and postempiricist philosophy of science in archaeology. J Archaeol Method Theory 3:1–30

Tsourlos PI, Tsokas GN (2011) Non-destructive electrical resistivity tomography survey at the south walls of the acropolis of Athens. Archaeol Prospect 18:173–186

Tuller H, Hofmeister U, Daley S (2008) Spatial analysis of mass grave mapping data to assist in the Reassociation of disarticulated and commingled human remains. In: Adams BJ, Byrd JE (eds) Recovery, analysis, and identification of commingled human remains. Humana Press, New York, pp 7–29

Ubelaker DH (2015) The global practice of forensic science. Wiley-Blackwell, Chichester

Ubelaker DH (2018) Species determination from fragmentary evidence. In: Latham K, Bartelink E, Finnegan M (eds) New perspectives in forensic human skeletal identification. Elsevier, London, pp 197–200

Ubelaker D, Zarenko K (2011) Adipocere: what is known after over two centuries of research. Forensic Sci Int Available at: http://www.sciencedirect.com/science/article/pii/S0379073810005256. Accessed 20 May 2016

Ubelaker DH, Lowenstein JM, Hood DG (2004) Use of solid-phase double-antibody radioimmunoassay to identify species from small skeletal fragments. J Forensic Sci 49:924–929

Ueland M, Breton H, Forbes S (2014) Bacterial populations associated with early-stage adipocere formation in lacustrine waters. Int J Legal Med http://link.springer.com/article/10.1007/s00414-013-0907-7. Accessed 20 May 2016

Ulguim PF (2017) Recording in situ human remains in three dimensions: applying digital image-based modeling. In: Errickson D, Thompson T (eds) Human remains: another dimension. Academic Press, London, pp 71–92

US Department of Commerce NO and AA. How far does light travel in the ocean? https://oceanservice.noaa.gov/facts/light_travel.html. Accessed 1 Mar 2018

van Daalen MA et al (2017) An aquatic decomposition scoring method to potentially predict the postmortem submersion interval of bodies recovered from the North Sea. J Forensic Sci 62(2):369–373. https://doi.org/10.1111/1556-4029.13258 Accessed 5 March 2017

Van Patten IT, Delhauer PQ (2007) Sexual homicide: a spatial analysis of 25 years of deaths in Los Angeles. J Forensic Sci 52:1129–1141

Vass A (2011) The elusive universal post-mortem interval formula. Forensic Sci Int Available at: http://www.sciencedirect.com/science/article/pii/S037907381000232X Accessed 20 May 2016

Verhagen P, Whitley TG (2012) Integrating archaeological theory and predictive modelling: a live report from the scene. J Archaeol Method Theory 19:49–100

Villa C, Gaudio D, Cattaneo C, Buckberry J, Wilson AS, Lynnerup N (2015) Surface curvature of pelvic joints from three laser scanners: separating anatomy from measurement error. J Forensic Sci 60:374–381

Vrijenhoek RC, Johnson SB, Rouse GW (2009) A remarkable diversity of bone-eating worms (Osedax; Siboglinidae; Annelida). BMC Biol 7(74). Available at: https://bmcbiol.biomedcentral.com/articles/10.1186/1741-7007-7-74). Accessed 26 July 2017

Walker RS (2012) UK Gemini: specification for discovery metadata for geospatial resources, v2.2. Association for Geographic Information

Warnasch SC (2016) Forensic archaeological recovery of a large-scale mass disaster scene: lessons learned from two complex recovery operations at the world trade center site. J Forensic Sci 61(3):584–593. https://doi.org/10.1111/1556-4029.13025

Wetherill C (1860) On Adipocire, and its formation. In: Transactions of the American Philosophical Society. Available at: http://www.jstor.org/stable/3231924 Accessed 23 Oct 2016

Widacki J (ed) (1983) Wykorzystanie kryminalistyki i medycyny sądowej w badaniach historycznych. Uniwersytet Śląski, Katowice

Widya M, Moffatt C, Simmons T (2012) The formation of early stage adipocere in submerged remains: a preliminary experimental study. J Forensic Sci Available at: http://onlinelibrary.wiley.com/doi/10.1111/j.1556-4029.2011.01980.x/full Accessed 20 May 2016

Wilhelmson H, Dell'Unto N (2015) Virtual taphonomy: a new method integrating excavation and postprocessing in an archaeological context. Am J Phys Anthropol 157(2):305–321

Wilkinson C (2004) Forensic facial reconstruction. Cambridge University Press, Cambridge

Wilkinson C, Rynn C (eds) (2012) Craniofacial identification. Cambridge University Press, Cambridge

William B (1995) Humam osteology: a laboratory and field manual. Missouri Archaeological Society

Wilson AS (2017) Taphonomic alterations to hair and nail. In: Schotsmans EMJ, Nicholas M, Forbes SL (eds) Taphonomy of human remains: forensic analysis of the dead and the depositional environment. Wiley, Hoboken, pp 81–91

Wilson AS, Janaway RC, Holland AD, Dodson HI, Baran E, Pollard AM, Tobin DJ (2007) Modelling the buried human body environment in upland climes using three contrasting field sites. Forensic Sci Int 169(1):6–18

Wilson AS, Holland AD, Sparrow T (2017) Laser scanning of skeletal pathological conditions. In: Errickson D, Thompson T (eds) Human remains: another dimension. Academic Press, London, pp 123–134

World Health Organization (2017). Drowning: Fact sheet. Available at: http://www.who.int/mediacentre/factsheets/fs347/en/. Accessed 5 Oct 2017

Wylie A (2002) Thinking from things: essays in the philosophy of archaeology. University of California Press, Berkley

Wylie A (2011) Critical distance: stabilizing evidential claims in archaeology. In: Dawid P, Twining W, Vasilaki M (eds) Evidence, inference and enquiry. Oxford University Press, Oxford, pp 371–394

Wylie MA (1989) The interpretative dilemma. In: Pinsky V, Wylie A (eds) Critical traditions in contemporary archaeology. Cambridge University Press, Cambridge, pp 18–27

Yeboah D, Taylor S, McPolin D, Gilfillan R (2013) Pull-out behaviour of axially loaded basalt fibre reinforced polymer rods bonded perpendicular to the grain of glulam elements. Constr Build Mater 38:962–969

Zakšek K, Oštir K, Kokalj Z (2011) Sky-view factor as a relief visualisation technique. Remote Sens 3:398–415

Zalewska AI (ed) (2016) Archeologia współczesności. Stowarzyszenie Naukowe Archeologów Polskich, Warszawa

Zimmerman K, Wallace J (2008) Estimating a postmortem submersion interval using algal diversity on mammalian carcasses in brackish marshes. J Forensic Sci Available at: https://scholar.google.com/scholar?start=30&q=postmortem+submersion+interval&hl=fr&as_sdt=0,5#5 Accessed 20 May 2016

Zlot R, Bosse M, Greenop K, Jarzab Z, Juckes E, Roberts J (2014) Efficiently capturing large, complex cultural heritage sites with a handheld mobile 3D laser mapping system. J Cult Herit 15(6):670–678

Index

P. M. Barone, W. J. M. Groen (eds.), *Multidisciplinary Approaches to Forensic Archaeology*, Soil Forensics, https://doi.org/10.1007/978-3-319-94397-8

MIX
Papier aus verantwortungsvollen Quellen
Paper from responsible sources
FSC® C105338

If you have any concerns about our products,
you can contact us on
ProductSafety@springernature.com

In case Publisher is established outside the EU,
the EU authorized representative is:
Springer Nature Customer Service Center GmbH
Europaplatz 3, 69115 Heidelberg, Germany

Printed by Libri Plureos GmbH
in Hamburg, Germany